Reproducible Finance with R

Code Flows and Shiny Apps for Portfolio Analysis

Chapman & Hall/CRC
The R Series

Series Editors

John M. Chambers, Department of Statistics Stanford University Stanford, California, USA
Torsten Hothorn, Division of Biostatistics University of Zurich Switzerland
Duncan Temple Lang, Department of Statistics University of California, Davis Davis, California, USA
Hadley Wickham, RStudio, Boston, Massachusetts, USA

For more information about this series, please visit: https://www.crcpress.com/go/the-r-series

Reproducible Finance
with R
Code Flows and Shiny Apps for Portfolio Analysis

Jonathan K. Regenstein, Jr.

CRC Press
Taylor & Francis Group
Boca Raton London New York

CRC Press is an imprint of the
Taylor & Francis Group, an **informa** business
A CHAPMAN & HALL BOOK

CRC Press
Taylor & Francis Group
6000 Broken Sound Parkway NW, Suite 300
Boca Raton, FL 33487-2742

Printed on acid-free paper
Version Date: 20180830

International Standard Book Number-13: 978-1-1384-8403-0 (Paperback)

Visit the Taylor & Francis Web site at
http://www.taylorandfrancis.com

and the CRC Press Web site at
http://www.crcpress.com

To Olivia, Roxanne and Eloisa

Contents

Preface

The Motivation

This book has two practical motivations and one philosophic motiavtion. The two practical movitations are: (1) to introduce R to finance professionals, or people who want to become finance professionals, who wish to move beyond Excel for their quantitative work and (2) to introduce various finance coding paradigms to R coders. The book seeks to be a resource for R coders interested in finance, or financiers who are interested in R or quantitative work generally.

The philosophical motivation for both audiences is to demonstrate and emphasize *reproducible* finance with good R code. The reason the first word in the title of this book is not 'financial', or 'quantitative', or 'algorithmic' is that the driving force behind this book is clean, neat, readable, reusable and reproducible R code for finance. We will prioritize code that is understandable over code that is theoretically brilliant.

Structure of the book

This book is organized into 4 sections: *Returns, Risk, Portfolio Theory* and *Practice and Applications*. Each of those sections contains several chapters with specific problems to solve. This book can be read sequentially or it can be accessed randomly. If you wished to read the CAPM chapter today and the skewness chapter tomorrow, that would work fine as each chapter is written to stand on its own.

More generally, this book is structured around building a portfolio and then analyzing that portfolio, or what I think of as telling the story of that portfolio with data. We will use that portfolio as the vehicle for exploring R and that is quite purposeful. I often hear the question asked, "I want to learn R, what steps should I take?". And if that question is posed to me, I like to ask, "What problem are you trying to solve?". The best way to start learning R or deepen your knowledge of R is to have a specific project or problem that you need to solve. Thus, the book is structured as the project of analysing our custom portfolio.

Who Should Read This Book

This book is for people who work in finance or want to work in finance, and specifically in portfolio management, and want to use the R programming language for quantitative work. If you have a CFA® or are working towards one, or are in a finance program in undergrad or an MBA program, or are an investment analyst at a bank, fund or investment manager, and you have an interest in R for data science or quant work, this book will introduce the landscape and get you started with practical, reproducible examples.

It will be very helpful to have some background in the R programming language *or* in finance *or* in statistics. The most important thing is a desire to learn about reproducible R code for finance.

For Readers Who are New to R

We are not going to cover the fundamentals or theory of R programming. We are going to cover the practical stuff you need to start producing impactful financial work with R.

There are a lot great of resources for learning R.

I highly recommend the book *R for Data Science*, which is available for free here:

http://r4ds.had.co.nz/

If you prefer videos, there is a free introduction to R here:

datacamp.com/courses/free-introduction-to-r

For Readers Who are New to Finance

Similar to our treatment of R, we will not delve deep into portfolio theory but we will cover many practical examples. For those readers who want more background, a good starting text is *Investments* by Bodie, Kane and Marcus.

Getting the Most ouf of this Book

This book is meant to be read alongside a computer that is connected to the internet so that you can see the Shiny apps online and grab code from the website as needed. It is hard to copy code out of a book.

To code along with this book, please load R and RStudio on to your computer.

To download R, go to:

`https://cloud.r-project.org`

and then click on the link for either Mac, Windows or Linux depending on your computer.

To install RStudio, go to:

`http://www.rstudio.com/download`

RStudio is an integrated development environment (or IDE) for R programming. It makes writing and running R code more fun.

If all of that is completely mystifying, have a look at this section from *R for Data Science*:

r4ds.had.co.nz/introduction.html#prerequisites

If you are 100% brand new to R, getting R and RStudio installed on your computer can be the hardest and most frustrating part. It's not intuitive and it has nothing to do with the work you want to do. I have been there and I have smashed my head in frustration many times. Stick with it and use the online resources for guidance. Once it's done, it's done. In terms of how to approach a getting the code for this book, at the end of each section I provide a link where you can find the bare code, with no explanatory text, that was covered in each chapter of that section. You can copy that bare code to your computer before reading the section and run the code alongside the book, or you can read the chapters first, and then copy and run the code to ensure it all still makes sense. Either way would work well, but I do recommend to run the code on your computer at some point.

Packages

R the programming language consists of base R and the packages that have been built on top of it. Once you have downloaded base R onto your computer and installed RStudio, you need to install the packages we will be using for this book.

To install a package on your computer, run install.packages("name of package"). To use that package, place library(name of package) at the top of your R script or RMarkdown file and run it.

Here are the commands to get the packages for this book:

```
# tidyverse contains the packages tidyr, ggplot2, dplyr,
# readr, purrr and tibble
install.packages("tidyverse")
install.packages("lubridate")
install.packages("readxl")
install.packages("highcharter")
install.packages("tidyquant")
install.packages("timetk")
install.packages("tibbletime")
install.packages("quantmod")
install.packages("PerformanceAnalytics")
install.packages("scales")


library(tidyverse)
library(lubridate)
library(readxl)
library(highcharter)
library(tidyquant)
library(timetk)
library(tibbletime)
library(quantmod)
library(PerformanceAnalytics)
library(scales)
```

Conventions

Package names, functions, and other R objects are in typewriter set (e.g., dplyr) and function names are followed by parentheses (e.g., median()).

Acknowledgments

A lot of people helped me when I was writing the book, by creating great R packages and/or offering guidance. The amazing thing about open source software is that many of these people do their work thanklessly and have no idea how much they help R users like me. Thank you to:

Garrett Grolemund, for help with coding questions and for writing *R for Data Science*.

Jenny Bryan, for guidance on purrr and data import.

Joe Rickert, for encouraging and publishing a first time writer on the Rviews blog.

Hadley Wickham, for creating the tidyverse and for writing *R for Data Science*

Joe Cheng, who created Shiny and opened the word of app development to R coders.

Mine Çetinkaya-Rundel, who helped me learn Shiny.

Yihui Xie, for creating blogdown, bookdown, and answering innumerable questions about RMarkdown.

Matt Dancho, for creating tidyquant, timetk, and tibbletime.

Rex Macey, for critical comments about the financial substance.

JJ Allaire, for creating RStudio and making R accessible to the world.

Dominic Hughes, the man who introduced me to R.

Josh Ulrich and Jeffrey Ryan, for all their work in the xts world.

About the Author

Jonathan Regenstein is the Director of Financial Services at RStudio. He studied international relations at Harvard University and did graduate work in political economy at Emory University before leaving the PhD program. He lives in Atlanta, GA.

1

Introduction

What is Reproducible Finance?

Reproducible finance is a philosophy about how to do quantitative, data science-driven financial analysis. The root of this philosophy is that the data and code that lead to a decision or conclusion should be able to be understood and then replicated in an efficient way. The code itself should tell a clear story when read by a human, just as it tells a clear story when read by a computer. This book applies the reproducible philosophy to R code for portfolio management.

That reproducible philosophy will manifest itself in how we tackle problems throughout this book. More specifically, instead of looking for the most clever code or smartest algorithm, this book prioritizes readable, reusable, reproducible work flows using a variety of R packages and functions. We will frequently solve problems in different ways, writing code from different packages and using different data structures to arrive at the exact same conclusion. To repeat, we will solve the same problems in a multitude of ways with different coding paradigms.

The motivation for this masochistic approach is for the reader to become comfortable working with different coding paradigms and data structures. Our goal is to be fluent or at least conversational to the point that we can collaborate with a variety of R coders, understand their work and make our work understandable to them. It's not enough that our work be reproducible for ourselves and other humans who possess our exact knowledge and skills. We want our work to be reproducible and reusable by a broad population of data scientists and quants.

Three Universes

This book focuses on three universes or paradigms for portfolio analysis with R. There are probably more than three fantastic paradigms but these are the three I encounter most frequently in industry.

xts

The first universe is what I call the xts world. xts is both a package and a type of object. xts stands for extensible time series. Most of our work in this book will be with time series, and indeed most financial work involves time series. An `xts` object is a matrix, that also, always, has a time index for the order of the data. It holds a time series, meaning it holds the observations and the times at which they occurred. An interesting feature of an `xts` object is that it holds dates in an `index` column. In fact that index column is considered column number zero, meaning it is not really a column at all. If we have an object called `financial_data` and wanted to access the dates, we would use `index(financial_data)`.

Why is the date index not given its own column? Because it is impossible to have an `xts` object but not have a date index. If the date index were its own column, that would imply that it could be deleted or removed.

In the `xts` world, there are two crucial packages that we will use: `quantmod` and `PerformanceAnalytics`. `quantmod` is how we will access the internet and pull in pricing data. That data will arrive to us formatted as an `xts` object.

`PerformanceAnalytics`, as the name implies, has several useful functions for analyzing portfolio performance in an `xts` object, such as `StdDev()`, `SharpeRatio()`, `SortinoRatio()`, `CAPM.Beta()`. We will make use of this package in virtually all of the chapters.

To learn more, have a look at the documentation:

cran.r-project.org/web/packages/PerformanceAnalytics/index.html

tidyverse

The second universe is known throughout the R community as the 'tidyverse'. The tidyverse is a collection of R packages for doing data science in a certain way. It is not specific to financial services and is not purpose built for time series analysis.

Within the tidyverse, we will make heavy use of the `dplyr` package for data wrangling, transformation and organizing. `dplyr` does not have built-in functions for our statistical calculations, but it does allow us to write our own functions or apply some other package's functions to our data.

In this world, our data will be in a data frame, also called a `tibble`. Throughout this book, I will use those two interchangeably: data frame = `tibble` in this book.

Why is it called the *tidy* verse? Because it expects and wants data to be tidy, which means:

```
(1) each variable has its own column
(2) each observation is a row
(3) each value is a cell
```

Learn more here:

tidyr.tidyverse.org/

We will explore how to make data tidy versus non-tidy throughout the book.

tidyquant

The third universe is `tidyquant`, which includes the `tidyquant`, `timetk` and `tibbletime` packages. This universe takes a lot of the best features of `xts`, `PerformanceAnalytics` and the tidyverse and lets them play well together. For example, `tidyquant` allows us to apply a function from `PerformanceAnalytics` to a tidy data frame, without having to convert it to an `xts` object.

Learn more here:

business-science.io/r-packages.html

Those three universes will provide the structure to our code as we work through calculations. As a result, where possible, each chapter or substantive task will follow a similar pattern: solve it via `xts`, solve it via `tidyverse`, solve it via `tidyquant` and verify that the results are the same. In this way, we will become familiar with data in different formats and using different paradigms.

For some readers, it might become tedious to solve each of our tasks in three different ways and if you decide you are interested in just one paradigm, feel free to read just that code flow for each chapter. The code flow for each universe can stand on its own.

Data Visualization

Data visualization is where we translate numbers into shapes and colors, and it will get a lot of attention in this book. We do this work so that humans who do not wish to dig into our data and code can still derive value from what we do. This human communication is how our quiet quantitative toiling becomes a transcendent revenue generator or alpha-producing strategy, Even if we plan to implement algorithms and never share our work outside of our own firm, the ability to explain and communicate is hugely important.

To the extent that clients, customers, partners, bosses, portfolio managers and anyone else want actionable insights from us, data visualizations will most certainly be more prominent in the discussion than the nitty gritty of code, data or even statistics. I will emphasize data visualization throughout the book and implore you to spend as much or more time on data visualizations as you do on the rest of quantitative finance.

When we visualize our results, object structure will again play a a role. We will generally chart `xts` objects using the `highcharter` package and tidy objects using the `ggplot2` package.

`highcharter` is an R package but `Highcharts` is a JavaScript library - the

R package is a hook into the JavaScript library. `Highcharts` is fantastic for visualizing time series and it comes with great built-in widgets for viewing different time frames. I highly recommend it for visualizing financial time series but you do need to buy a license to use it in a commercial setting.

Learn more at:

www.highcharts.com and

cran.r-project.org/web/packages/highcharter/highcharter.pdf

`ggplot2` is itself part of the tidyverse and as such it works best when data is tidy (we will cover what that word 'tidy' means when applied to a data object). It is one of the most popular data visualization packages in the R world.

Learn more at:

ggplot2.tidyverse.org/

Shiny Applications

Each of our chapters will conclude with the building of a Shiny application, so that by book's end, you will have the tools to build a suite of Shiny apps and dashboards for portfolio analysis. What is Shiny?

Shiny is an R package that was created by Joe Cheng. It wraps R code into interactive web applications so R coders do *not* need to learn HTML, CSS or JavaScript.

Shiny applications are immeasurably useful for sharing our work with end users who might not want to read code or open an IDE. For example, a portfolio manager might want to build a portfolio and see how a dollar would have grown in that portfolio or how volatility has changed over time, but he or she does not want to see the code, data and functions used for the calculation. Or, another PM might love the work we did on Portfolio 1, and have a desire to apply that work to Portfolios 2, 3 and 4 but under different economic assumptions.

Shiny allows that PM to change input parameters on the fly, run R code under the hood for new analytic results (without knowing its R code), and build new data visualizations.

After completing this book you will be able to build several portfolio management-focused Shiny apps. You will not be an expert on the theory that underlies Shiny or its reactive framework, but you will have the practical knowledge to code functional and useful apps.

We will build the following Shiny applications:

1) Portfolio Returns

2) Portfolio Standard Deviation

3) Skewness and Kurtosis of Returns

4) Sharpe Ratio

5) CAPM Beta

6) Fama-French Factor Model

7) Asset Contribution to Portfolio Standard Deviation

8) Monte Carlo Simulation

You can see all of those applications at the *Reproducible Finance* website:

www.reproduciblefinance.com/shiny

The full source code for every app is also available at that site. It is not necessary to view the apps live on the internet, but doing so will make it easier to understand what the code is doing.

Packages

The following are the packages that we will be using in this book.

To install a package on your computer, run `install.packages("name of package")`. To use that package, `library(name of package)'` at the top of your R script or RMarkdown file.

```r
# tidyverse contains the packages tidyr, ggplot2, dplyr,
# readr, purrr and tibble
install.packages("tidyverse")
install.packages("lubridate")
install.packages("readxl")
install.packages("highcharter")
install.packages("tidyquant")
install.packages("timetk")
install.packages("tibbletime")
install.packages("quantmod")
install.packages("PerformanceAnalytics")
install.packages("scales")


library(tidyverse)
library(lubridate)
library(readxl)
library(highcharter)
library(tidyquant)
```

```
library(timetk)
library(tibbletime)
library(quantmod)
library(PerformanceAnalytics)
library(scales)
```

A Note on Style

This book began as a monthly series of blog posts about R and finance. As a result, the tone and structure are decidedly blog-like. Plus, there is a lot of code in this book and dense prose coupled with code can be hard to digest. I prefer casual prose to accompany code. Please have a look at some of the posts on the book's website to see if this tone and structure appeal to you.

A Final Caveat: It's Alive!

One of R's most powerful traits is its collection of packages. It allows us to harness the past work and ingenuity of thousands of smart R coders. There are over 17,000 R packages as of the time I am typing this. As those packages evolve, our code might need to evolve as well. For example, the tibbletime package got a fantastic upgrade about one week before the first draft of this book and I added new use cases to explore those.

The flip side is that packages could change the day *after* this book or any of our code is published. If you upgrade a package to the newest version, there's a chance your code will need to be tweaked; if you don't upgrade a package, there's a chance your code is now outdated. It's an exciting and challenging aspect of being an R coder.

For that reason, this book is a living and breathing project, just like the R universe itself. If an amazing package update arrives to R in 6 months, I will code examples using the data in this book and post it on www.reproduciblefinance.com/code. If I see an interesting way to apply or extend the code flows in this book to a different use case, I will likewise post it to www.reproduciblefinance.com/code.

Enough set up, let's get to it!

Returns

Welcome to our section on asset returns, wherein we perform the un-glamorous work of taking raw price data for 5 individual assets and transforming them into monthly returns for a single portfolio.

Our portfolio will consist of the following Exchange Traded Funds (ETFs):

+ SPY (S&P500 ETF) weighted 25%
+ EFA (a non-US equities ETF) weighted 25%
+ IJS (a small-cap value ETF) weighted 20%
+ EEM (an emerging-markets ETF) weighted 20%
+ AGG (a bond ETF) weighted 10%

As noted in the *Introduction*, we will be reviewing three coding paradigms in each chapter and, thus, by the end of this section we will have data objects from the xts world, the tidyverse, and the tidyquant world. Each of those objects will hold portfolio monthly returns from January 31, 2013 through December 31, 2017 and we will be working with those three objects for the remainder of the book.

To move from an empty R environment to one with three portfolio returns objects, we will take these steps:

1) Import daily prices from the internet, a csv file or xls file
2) Transform daily prices to monthly prices
3) Transform monthly prices to monthly returns
4) Visualize monthly returns
5) Calculate portfolio monthly returns based on asset monthly returns and weights
6) Visualize portfolio returns
7) Save the data objects for use throughout this book

To map a data science work flow onto this section, those steps encompass data import, cleaning, wrangling, transformation and initial visualization to make sure the wrangling has gone how we wish. Even though the substantive issues are not complex, we will painstakingly review the code to ensure that the data provenance is clear, reproducible and reusable. In fact, we will devote as much time to this section as we do to any of the more analytic sections. That might seem a bit unbalanced - after all, quants do not get paid to import, clean and wrangle data. But this work is fundamental to our more complex alpha-generating and risk-minimizing tasks. Our partners, collaborators and future selves will thank us for this effort when they want to update our models, extend our work or stress test our portfolios.

2

Asset Prices to Returns

Let's get to step 1 wherein we import daily price data for the 5 ETFs and save them to an xts object called prices. We will cover how to load this data from the internet, and then how to load it from a csv or xls file.

To get the data from the internet, we first choose ticker symbols and store them in a vector called symbols.

We do that with symbols <- c("SPY","EFA", "IJS", "EEM","AGG"). Those are the tickers for the 5 assets in our portfolio. If you want to change to different assets for testing, change those tickers. Note that if you wish to choose different stock tickers or create a different portfolio, you change the tickers in the symbols vector.

```
symbols <- c("SPY","EFA", "IJS", "EEM","AGG")
```

We then pass symbols to Yahoo! Finance via the getSymbols() function from the quantmod package. This will return an object with the opening price, closing price, adjusted price, daily high, daily low and daily volume for each ticker.

Note that we are enforcing a starting date of "2012-12-31" and an end date of "2017-12-31". That means we will be working with 5 years of data (when we convert to monthly returns, we will lose December of 2012). If you wish to pull data that is up-to-date as of today, you can remove the argument to = "2017-12-31" but then your raw data will be different from what is being used in this book.

To isolate the adjusted price, we use the map() function from the purrr package and apply Ad(get(.)) to the imported prices. This will get() the adjusted price from each of our individual price series. If we wanted the closing price, we would run Cl(get(.)). That . refers to our initial object.

We could stop here and have the right substance - daily prices for 5 ETFs - but the format would not be great as we would have a list of 5 xts objects. This is because the map() function returns a list by default.

The reduce(merge) function will merge the 5 lists into one xts object. The merge() function looks for the date index shared by our objects and uses that index to align the data.

Finally, we want intuitive column names and use `colnames<-` to rename the columns according to the `symbols` object.

```
prices <-
  getSymbols(symbols,
             src = 'yahoo',
             from = "2012-12-31",
             to = "2017-12-31",
             auto.assign = TRUE,
             warnings = FALSE) %>%
  map(~Ad(get(.))) %>%
  reduce(merge) %>%
  `colnames<-`(symbols)
```

Note that we are sourcing data from Yahoo! Finance with `src = 'yahoo'`. In industry, we almost certainly would not be pulling from the internet but instead would be accessing an internal database, or loading a csv or xls file that someone had made available.

To prepare for that inevitability, let's look at how to load a csv file into RStudio.

First, navigate to this URL:

www.reproduciblefinance.com/data/data-download/

and click on 'CSV'.

After clicking that button, there should be a downloaded file on your computer called `Data.csv`. On my machine, the path to that csv file is `/Users/myname/Downloads/Data.csv`.

To load it into RStudio, we use the `read_csv()` function from `readr`. Note that `read_csv()` would interpret our `date` column as a numeric but we can specify that our `date` column is in year-month-day format. We do that with `col_types = cols(date = col_date(format = "%Y-%m-%d"))`.

```
prices <-
  read_csv("path to your data.csv",
    col_types =
      cols(date =
             col_date(format = "%Y-%m-%d"))) %>%
  tk_xts(date_var = date)
```

The function `tk_xts(date_var = date)` converts the data frame to an `xts` object. Look back at the prices object we created with `getSymbols()` and recall that is an `xts` object as well. That is why we include the call to `tk_xts()`, to be consistent with our previous import.

Look again at:

www.reproduciblefinance.com/data/data-download/

There is the option to download an xls file by clicking the `Excel` button, which downloads an xls file with the path `/Users/myname/Downloads/Data.xls`.

We import it to RStudio using the `read_excel()` function from the `readxl` package.

```
prices <-
  read_excel("path to your excel spreadsheet",
             col_types = c("text", "numeric",
                           "numeric", "numeric",
                           "numeric", "numeric")) %>%
  mutate(date = ymd(date)) %>%
  tk_xts(date_var = date)
```

Notice that we again had to deal with the `date` column and coerced it to a date format using the `ymd()` function from the `lubdridate` package and the `mutate()` function from `dplyr`. We often need to coerce our data a bit upon import.

We have now covered three methods for importing the raw data needed for this book - from the internet, from csv and from xls - and we devoted significant time to this task for three reasons.

First, we added a few data import functions to our toolkit. Those functions make the beginning of a project smooth, especially for new R coders who want to get to work and not get mired in data import.

Second, we emphasized an important but un-glamorous best practice: make raw data accessible. If a collaborator needs to use a function or package to get our raw data, we need to document it and explain it. That can be boring and it is not generally the main job of a finance quant or portfolio manager. It is, though, an important responsibility in the reproducible work flow. And remember that most important of collaborators: yourself in 6 months when it's time to update or refactor your model. That future self might not remember where to find this data unless it is clearly delineated.

Third, by documenting and explaining the R functions used to import the data, we also discussed the lineage or provenance of our data. Here it originated on Yahoo! Finance and still lives there, but it also lives in csv and xls files at this book's website. It would have been hard to document how to import this data without also explaining where the data resides and from whence it originated. For ETF prices, that is not a major concern. However, if we were using alternative, custom or proprietary data, it would be crucial to explain and document the data lineage so that our collaborators could understand where the data originated, where it resides now and how it can be used to reproduce or expand upon our work.

With all that said, have one final look at the `prices` object.

```
head(prices, 3)
```

```
              SPY   EFA    IJS    EEM   AGG
2012-12-31 127.7 48.21 75.07 39.63 98.20
2013-01-02 131.0 48.95 77.13 40.41 98.08
2013-01-03 130.7 48.48 77.02 40.12 97.83
```

2.1 Converting Daily Prices to Monthly Returns in the xts world

Next we will convert daily prices to monthly log returns.

I mentioned in the introduction that we would be working in three universes - xts, `tidyverse` and `tidyquant` - the `prices` object is an `xts`, so we will start there.

The first observation in our `prices` object is December 31, 2012 (the last trading day of that year) and we have daily prices. We want to convert to those daily prices to monthly log returns based on the last reading of each month.

We will use `to.monthly(prices, indexAt = "last", OHLC = FALSE)` from the `quantmod` package. The argument `indexAt = "lastof"` tells the function whether we want to index to the first day of the month or the last day. If we wanted to use the first day, we would change it to `indexAt = "firstof"`.

```
prices_monthly <- to.monthly(prices,
                             indexAt = "lastof",
                             OHLC = FALSE)
```

```
head(prices_monthly, 3)
```

```
              SPY   EFA    IJS    EEM   AGG
2012-12-31 127.7 48.21 75.07 39.63 98.20
2013-01-31 134.3 50.00 79.08 39.52 97.59
2013-02-28 136.0 49.36 80.37 38.61 98.16
```

We have moved from an `xts` object of daily prices to an `xts` object of monthly prices. Note that we now have one reading per month, for the last day of each month.

Now we call `Return.calculate(prices_monthly, method = "log")` to convert to returns and save as an object called `asset_returns_xts`. Note this will give us log returns by the `method = "log"` argument. We could have used `method = "discrete"` to get simple returns.

```
asset_returns_xts <-
  Return.calculate(prices_monthly,
                   method = "log") %>%
  na.omit()

head(asset_returns_xts, 3)
```

```
                SPY      EFA      IJS       EEM
2013-01-31  0.04992  0.03661  0.05213  -0.002935
2013-02-28  0.01268 -0.01297  0.01618  -0.023105
2013-03-31  0.03727  0.01297  0.04026  -0.010235
                AGG
2013-01-31  -0.0062315
2013-02-28   0.0058912
2013-03-31   0.0009848
```

Notice in particular the date of the first value. We imported prices starting at "2012-12-31" yet our first monthly return is for "2013-01-31". This is because we used the argument `indexAt = "lastof"` when we cast to a monthly periodicity (try changing to `indexAt = "firstof"` and see the result). That is not necessarily good or bad, but it might matter if that first month's returns makes a difference in our analysis.

More broadly, this is a good time to note how our decisions in data transformation can affect the data that ultimately survive to our analytic stage. We just lost the first month of daily prices. Maybe that does not matter to us, or maybe we want to change the original start date to make sure we capture December of 2012. It depends on the task and the data, and if we outsource the data preparation, these decisions will be made before we even see the data.

From a substantive perspective, we have now imported daily prices, trimmed to adjusted prices, moved to monthly prices and transformed to monthly log returns, all in the `xts` world.

2.2 Converting Daily Prices to Monthly Returns in the tidyverse

We now take the same raw `prices` object and convert it to monthly returns using the tidyverse. We are leaving the `xts` structure and converting to a data frame. There are several differences between an `xts` object and a `tibble` but a very important one is the date, an essential component of most financial analysis. `xts` objects have a date *index*. In contrast, data frames have a date *column*, and that column is not necessary to the existence of the data frame. We could create a data frame that does not have a date column, for example, to hold cross-sectional data. An `xts` object, by definition, has a date index.

Our conversion from `xts` to a `tibble` starts with `data.frame(date = index(.))`, which (i) coerces our object into a data frame and (ii) adds a date column based on the index with `date = index(.)`. The date index of the `xts` object is preserved as row names and must be explicitly removed with `remove_rownames()`.

Next we turn to the `gather()` function from the `tidyr` package to convert our new data frame into long format. We have not done any calculations yet, we have only shifted from wide format, to long, tidy format.

Next, we want to calculate log returns and add those returns to the data frame. We will use `mutate` and our own calculation to get log returns: `mutate(returns = (log(prices) - log(lag(prices))))`. We now have log returns and will not need the raw prices data. It can be removed with `select(-prices)`. The `select()` function will select columns to keep, but if we had a negative sign, it will remove a column.

Our last two steps are to `spread` the data back to wide format, which makes it easier to compare to the `xts` object and easier to read, but is not a best practice in the tidyverse. We are going to look at this new object and compare to the `xts` object above, so we will stick with wide format for now.

Finally, we want to reorder the columns to align with how we first imported the data using the `symbols` vector - the first line of code we ran in this chapter.

```
asset_returns_dplyr_byhand <-
  prices %>%
  to.monthly(indexAt = "lastof", OHLC = FALSE) %>%
  # convert the index to a date
  data.frame(date = index(.)) %>%
  # now remove the index because it got converted to row names
  remove_rownames() %>%
```

```
  gather(asset, prices, -date) %>%
  group_by(asset) %>%
  mutate(returns = (log(prices) - log(lag(prices)))) %>%
  select(-prices) %>%
  spread(asset, returns) %>%
  select(date, symbols)
```

The above code chunk contains quite a few steps. A good way to work through them is in pieces. Try commenting all but 3 lines of code and see the result, then add back in a line or two, and see the result, etc.

Finally, have a quick peek at the new object.

```
head(asset_returns_dplyr_byhand, 3)
```

```
# A tibble: 3 x 6
  date            SPY      EFA      IJS       EEM
  <date>         <dbl>    <dbl>    <dbl>     <dbl>
1 2012-12-31 NA          NA       NA       NA
2 2013-01-31  0.0499    0.0366   0.0521   -0.00294
3 2013-02-28  0.0127   -0.0130   0.0162   -0.0231
# ... with 1 more variable: AGG <dbl>
```

Notice that our object now includes an NA reading for December 2012, whereas xts excluded it altogether. Remove the first row with the na.omit() function.

```
asset_returns_dplyr_byhand <-
  asset_returns_dplyr_byhand %>%
  na.omit()
```

Our two objects are now consistent and we have a tibble of portfolio monthly returns.

2.3 Converting Daily Prices to Monthly Returns in the tidyquant world

Let's explore the tidyquant paradigm for converting to log returns. We will start with the very useful tk_tbl() function from the timetk package.

In the piped workflow below, our call to tk_tbl(preserve_index = TRUE, rename_index = "date") (1) converts prices from xts to tibble, (2) converts the date *index* to a date *column*, and (3) renames it as "date" (since the index is being converted, it does not need to be removed as we did above).

Next, instead of using to.monthly and mutate, and then supplying our own calculation, we use tq_transmute(mutate_fun = periodReturn, period = "monthly", type = "log") and go straight from daily prices to monthly log returns.

```
asset_returns_tq_builtin <-
  prices %>%
  tk_tbl(preserve_index = TRUE,
         rename_index = "date") %>%
  gather(asset, prices, -date) %>%
  group_by(asset) %>%
  tq_transmute(mutate_fun = periodReturn,
               period = "monthly",
               type = "log") %>%
  spread(asset, monthly.returns) %>%
  select(date, symbols) %>%
  slice(-1)

head(asset_returns_tq_builtin, 3)
```

```
# A tibble: 3 x 6
  date          SPY     EFA     IJS      EEM       AGG
  <date>       <dbl>   <dbl>   <dbl>    <dbl>     <dbl>
1 2013-01-31 0.0499  0.0366  0.0521 -0.00294 -0.00623
2 2013-02-28 0.0127 -0.0130  0.0162 -0.0231   0.00589
3 2013-03-28 0.0373  0.0130  0.0403 -0.0102   0.000985
```

Once again, we needed to remove the first row and did so with slice(-1), which removed the first row (if we had used slice(1), we would have *kept* just the first row).

That tidyquant method produced the same output as the tidyverse method - a tibble of monthly log returns.

2.4   Converting Daily Prices to Monthly Returns with tibbletime

This is a good time to introduce the relatively new tibbletime package, which is purpose-built for working with time-aware tibbles. In the flow below, we will first convert `prices` to a `tibble` with `tk_tbl()`. Then, we convert to a `tibbletime` object with a `as_tbl_time(index = date)` and then convert to monthly prices with `as_period(period = "month", side = "end")`. The side argument anchors to the end of the month instead of the beginning. Try changing it to `side = "start"`.

```
asset_returns_tbltime <-
  prices %>%
  tk_tbl(preserve_index = TRUE,
         rename_index = "date") %>%
  # this is the the tibbletime function
  as_tbl_time(index = date) %>%
  as_period(period = "month",
            side = "end") %>%
  gather(asset, returns, -date) %>%
  group_by(asset) %>%
  tq_transmute(mutate_fun = periodReturn,
               type = "log") %>%
  spread(asset, monthly.returns) %>%
  select(date, symbols) %>%
  slice(-1)
```

This flow might not seem efficient - going from `xts` to `tibble` to `tibbletime` - but in future chapters we will see that rolling functions are smoother with `rollify()` and we can absorb some inefficiency now for future gains. Plus, the package is new and its capabilities are growing fast.

Before we move on, a quick review of our 4 monthly log return objects:

```
head(asset_returns_xts, 3)
```

```
                SPY       EFA      IJS       EEM
2013-01-31  0.04992   0.03661  0.05213  -0.002935
2013-02-28  0.01268  -0.01297  0.01618  -0.023105
2013-03-31  0.03727   0.01297  0.04026  -0.010235
                AGG
2013-01-31  -0.0062315
2013-02-28   0.0058912
2013-03-31   0.0009848
```

```
head(asset_returns_dplyr_byhand, 3)
```

```
# A tibble: 3 x 6
   date        SPY     EFA     IJS      EEM      AGG
```

```
     <date>        <dbl>    <dbl>  <dbl>    <dbl>       <dbl>
1 2013-01-31 0.0499    0.0366 0.0521 -0.00294 -0.00623
2 2013-02-28 0.0127   -0.0130 0.0162 -0.0231   0.00589
3 2013-03-31 0.0373    0.0130 0.0403 -0.0102   0.000985
```

```
head(asset_returns_tq_builtin, 3)
```

```
# A tibble: 3 x 6
    date          SPY      EFA    IJS      EEM        AGG
    <date>        <dbl>    <dbl>  <dbl>    <dbl>      <dbl>
1 2013-01-31 0.0499    0.0366 0.0521 -0.00294 -0.00623
2 2013-02-28 0.0127   -0.0130 0.0162 -0.0231   0.00589
3 2013-03-28 0.0373    0.0130 0.0403 -0.0102   0.000985
```

```
head(asset_returns_tbltime, 3)
```

```
# A time tibble: 3 x 6
# Index: date
    date          SPY      EFA    IJS      EEM        AGG
    <date>        <dbl>    <dbl>  <dbl>    <dbl>      <dbl>
1 2013-01-31 0.0499    0.0366 0.0521 -0.00294 -0.00623
2 2013-02-28 0.0127   -0.0130 0.0162 -0.0231   0.00589
3 2013-03-28 0.0373    0.0130 0.0403 -0.0102   0.000985
```

Do we notice anything of interest?

First, look at the date in each object. `asset_returns_xts` has a date index, not a column. That index does not have a name. It is accessed via `index(asset_returns_xts)`.

The `tibbles` have a column called "date", accessed via the `$date` convention, e.g. `asset_returns_dplyr_byhand$date`. That distinction is not important when we read with our eyes, but it is very important when we pass these objects to functions.

Second, each of these objects is in "wide" format, which in this case means there is a column for each of our assets: SPY has a column, EFA has a column, IJS has a column, EEM has a column, AGG has a column.

This is the format that `xts` likes and this format is easier for a human to read. However, the tidyverse calls for this data to be in long or tidy format where each variable has its own column. For asset returns to be tidy, we need a column called "date", a column called "asset" and a column called "returns".

To see that in action, here is how it looks.

```
asset_returns_long <-
  asset_returns_dplyr_byhand %>%
  gather(asset, returns, -date) %>%
  group_by(asset)

head(asset_returns_long, 3)
```

```
# A tibble: 3 x 3
# Groups:    asset [1]
  date         asset returns
  <date>       <chr>   <dbl>
1 2013-01-31 SPY     0.0499
2 2013-02-28 SPY     0.0127
3 2013-03-31 SPY     0.0373
```

```
tail(asset_returns_long, 3)
```

```
# A tibble: 3 x 3
# Groups:    asset [1]
  date         asset    returns
  <date>       <chr>      <dbl>
1 2017-10-31 AGG      0.000978
2 2017-11-30 AGG     -0.00148
3 2017-12-31 AGG      0.00441
```

`asset_returns_long` has 3 columns, one for each variable: date, asset, return. This format is harder for a human to read - we can see only the first several readings for *one* asset. From a tidyverse perspective, this is considered 'tidy' and it is the preferred format.

Before we move on, look at `asset_returns_xts` and our various wide `tibbles`, then look at the long, tidy object `asset_returns_long` object. Be sure the logic of how we got from daily prices to log returns makes sense for each code flow.

2.5   Visualizing Asset Returns in the xts world

It might seem odd that visualization is part of the data import and wrangling work flow and it does not have to be: we could jump straight into the process of converting these assets into a portfolio. However, it is a good practice to chart individual returns because once a portfolio is built, we are unlikely to back track to visualizing on an individual basis. Yet, those individual returns are the building blocks and raw material of our portfolio and visualizing them is a great way to understand them deeply. It also presents an opportunity to look for outliers, or errors, or anything unusual to be corrected before we move too far along in our analysis.

For the purposes of visualizing returns, we will work with two of our monthly log returns objects, `asset_returns_xts` and `asset_returns_long` (the tidy, long-formatted `tibble`).

We start with the `highcharter` package to visualize the `xts` formatted returns.

`highcharter` is an R package but `Highcharts` is a JavaScript library. The R package is a hook into the JavaScript library. `Highcharts` is fantastic for visualizing time series and it comes with great built-in widgets for viewing different time frames, plus we get to use the power of JavaScript without leaving the world of R code.

Not only are the visualizations nice, but `highcharter` "just works" with `xts` objects in the sense that it reads the index as dates without needing to be told. We pass in an `xts` object and let the package do the rest. I highly recommend it for visualizing financial time series but you do need to buy a license for use in a commercial setting.[1]

Let's see how it works for charting our asset monthly returns.

First, we set `highchart(type = "stock")` to get a nice line format that was purpose-built for stocks.

Then we add each of our series to the `highcharter` code flow with `hc_add_series(asset_returns_xts[, symbols[1]], name = symbols[1])`. Notice that we can use our original `symbols` object to reference the columns. This will allow the code to run should we change to different ticker symbols at the outset.

```
highchart(type = "stock") %>%
  hc_title(text = "Monthly Log Returns") %>%
  hc_add_series(asset_returns_xts[, symbols[1]],
                name = symbols[1]) %>%
  hc_add_series(asset_returns_xts[, symbols[2]],
                name = symbols[2]) %>%
  hc_add_series(asset_returns_xts[, symbols[3]],
                name = symbols[3]) %>%
  hc_add_series(asset_returns_xts[, symbols[4]],
                name = symbols[4]) %>%
```

---

[1] For more details, see www.highcharts.com.

```
hc_add_series(asset_returns_xts[, symbols[5]],
                name = symbols[5]) %>%
hc_add_theme(hc_theme_flat()) %>%
hc_navigator(enabled = FALSE) %>%
hc_scrollbar(enabled = FALSE) %>%
hc_exporting(enabled = TRUE) %>%
hc_legend(enabled = TRUE)
```

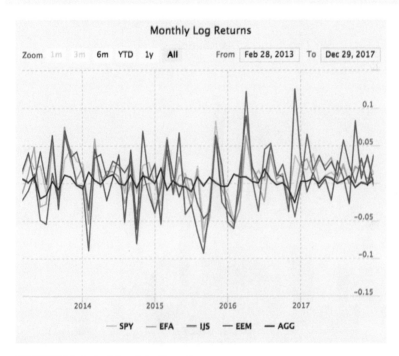

FIGURE 2.1: Monthly Log Returns

Take a look at figure 2.1 It has a line for the monthly log returns of each our ETFs (and in my opinion it's starting to get crowded).

Highcharter also has the capacity for histogram making. One method is to first call the base function `hist` on the data along with the arguments for breaks and `plot = FALSE`. Then we can call `hchart` on that object. Look at Figure 2.2 to see the results.

```
hc_hist <- hist(asset_returns_xts[, symbols[1]],
                breaks = 50,
                plot = FALSE)
```

```
hchart(hc_hist, color = "cornflowerblue") %>%
  hc_title(text =
             paste(symbols[1],
                   "Log Returns Distribution",
                   sep = " ")) %>%
  hc_add_theme(hc_theme_flat()) %>%
  hc_exporting(enabled = TRUE) %>%
  hc_legend(enabled = FALSE)
```

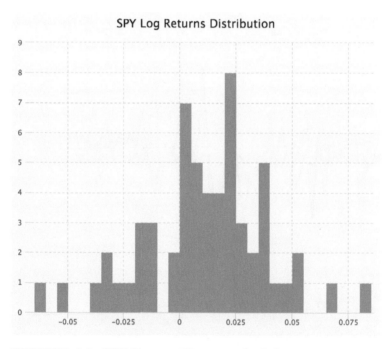

FIGURE 2.2: SPY Returns Histogram highcharter

Now we could create a function for building those histograms with the following
code:

```
hc_hist_fun <- function(n = 1, object, color){
  hc_hist <- hist(object[, symbols[n]],
                  breaks = 50,
                  plot = FALSE)

hchart(hc_hist, color = color) %>%
  hc_title(text =
             paste(symbols[n],
```

```
                    "Log Returns Distribution",
                    sep = " ")) %>%
  hc_add_theme(hc_theme_flat()) %>%
  hc_exporting(enabled = TRUE) %>%
  hc_legend(enabled = FALSE)
}
```

Then we could create 5 histograms for our assets, each with a different color by calling that function 5 times.

```
# Not run

hc_hist_fun(1, asset_returns_xts, "cornflowerblue")
hc_hist_fun(2, asset_returns_xts, "green")
hc_hist_fun(3, asset_returns_xts, "pink")
hc_hist_fun(4, asset_returns_xts, "purple")
hc_hist_fun(5, asset_returns_xts, "yellow")
```

We could also use map() from the purrr package to apply that function to our xts object, looping over each column to create a separate histogram for each.

```
# Not run
map(1:5, hc_hist_fun, asset_returns_xts, "blue")
```

Either of those flows works fine, but we had to do the work.

Let's head to the tidyverse and explore another visualization flow.

2.6 Visualizing Asset Returns in the tidyverse

ggplot2 is a very widely-used and flexible visualization package, and it is part of the tidyverse. We will use it to build a histogram and have our first look at how tidy data plays nicely with functions in the tidyverse.

In the code chunk below, we start with our tidy object of returns, asset_returns_long, and then pipe to ggplot() with the %>% operator. Next, we call ggplot(aes(x = returns, fill = asset)) to indicate that returns will be on the x-axis and that the bars should be filled with a different color for each asset. If we were to stop here, ggplot() would build an empty chart and that is because we have told it that we want a chart with certain x-axis values, but we have not told it what kind of chart to build. In `ggplot()` parlance, we have not yet specified a `geom`.

We use `geom_histogram()` to build a histogram and that means we do not specify a y-axis value, because the histogram will be based on counts of the returns.

Because the data frame is tidy and grouped by the `asset` column (recall when it was built we called `group_by(asset)`), `ggplot()` knows to chart a separate histogram for each asset. `ggplot()` will automatically include a legend since we included `fill = asset` in the `aes()` call.

```
asset_returns_long %>%
    ggplot(aes(x = returns, fill = asset)) +
    geom_histogram(alpha = 0.45, binwidth = .005) +
    ggtitle("Monthly Returns Since 2013")
```

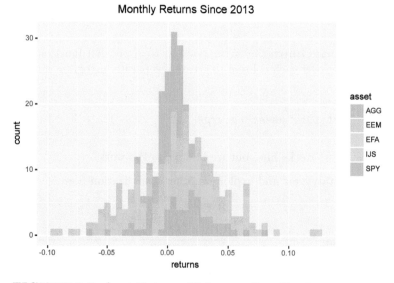

FIGURE 2.3: Asset Returns Histogram One Chart

Figure 2.3 shows different colors for each asset, but we did not have add each asset separately.

`facet_wrap(~asset)` will break this into 5 charts based on the `asset`, as shown in Figure 2.4.

```
asset_returns_long %>%
    ggplot(aes(x = returns, fill = asset)) +
    geom_histogram(alpha = 0.45, binwidth = .01) +
```

```
facet_wrap(~asset) +
ggtitle("Monthly Returns Since 2013") +
theme_update(plot.title = element_text(hjust = 0.5))
```

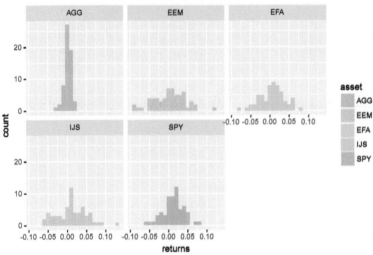

FIGURE 2.4: Asset Returns Histogram Faceted

Maybe we prefer a density line to visualize distributions. We can use `geom_density(alpha = 1)`, where the `alpha` argument is selecting a line thickness. We also add a label to the x and y axis with the `xlab` and `ylab` functions.

```
asset_returns_long %>%
  ggplot(aes(x = returns, colour = asset)) +
  geom_density(alpha = 1) +
  ggtitle("Monthly Returns Density Since 2013") +
  xlab("monthly returns") +
  ylab("distribution")  +
  theme_update(plot.title = element_text(hjust = 0.5))
```

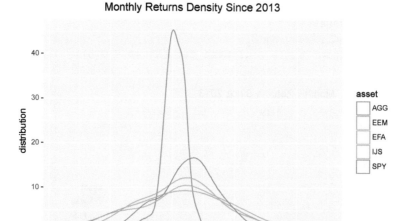

FIGURE 2.5: Asset Returns Density

Figure 2.5 shows density charts by asset and we have now made histograms and density plots. Perhaps we would like to combine both of those into one chart. `ggplot()` works in aesthetic layers, which means we can chart a histogram in one layer, and then add a layer with a density chart. Figure 2.6 shows the results when we start with a density chart for each asset, then layer on a histogram, then `facet_wrap()` by asset.

```
asset_returns_long %>%
  ggplot(aes(x = returns)) +
  geom_density(aes(color = asset), alpha = 1) +
  geom_histogram(aes(fill = asset), alpha = 0.45, binwidth = .01) +
  guides(fill = FALSE) +
  facet_wrap(~asset) +
  ggtitle("Monthly Returns Since 2013") +
  xlab("monthly returns") +
  ylab("distribution") +
  theme_update(plot.title = element_text(hjust = 0.5))
```

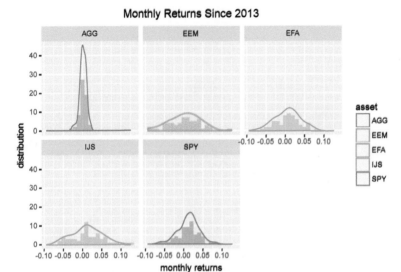

FIGURE 2.6: Asset Returns Histogram and Density Chart

Figure 2.6 is a good example of one chart, with histograms and line densities broken out for each of our assets. This would scale nicely if we had 50 assets and wanted to peek at more distributions of returns because `ggplot()` would recognize that each asset is a `group`. It would still be one call to `ggplot()` instead of 50.

3

Building a Portfolio

We spent a lot of time on the individual assets to make sure we had a good grasp of our data building blocks.

Next, we collect individual returns into a portfolio, which is a weighted set of asset returns. Accordingly, the first thing we need to do is assign a weight to each asset. Recall that our portfolio will have the following asset mix:

```
+ SPY (S&P500 ETF) weighted 25%
+ EFA (a non-US equities ETF) weighted 25%
+ IJS (a small-cap value ETF) weighted 20%
+ EEM (an emerging-markets ETF) weighted 20%
+ AGG (a bond ETF) weighted 10%
```

We need to create a weights vector that aligns with those allocations.

```
w <- c(0.25,
       0.25,
       0.20,
       0.20,
       0.10)
```

Before we use the weights in our calculations, have a quick sanity check in the next code chunk to make sure the weights and assets align.

```
tibble(w, symbols)
```

```
# A tibble: 5 x 2
      w symbols
  <dbl> <chr>
1 0.250 SPY
2 0.250 EFA
3 0.200 IJS
4 0.200 EEM
5 0.100 AGG
```

Finally, make sure the weights sum to 100%. We can eyeball this with 5 assets, but with 50 assets it would be better to run the sanity check to catch any errors as soon as possible

```
tibble(w, symbols) %>%
  summarise(total_weight = sum(w))
```

```
# A tibble: 1 x 1
  total_weight
         <dbl>
1           1.
```

Now we use those weights to convert the returns of 5 assets to the returns of 1 portfolio.

The return of a multi-asset portfolio is equal to the sum of the weighted returns of each asset.

We can implement that by assigning weights to variables according to our weights vector `w`.

```
w_1 <- w[1]
w_2 <- w[2]
w_3 <- w[3]
w_4 <- w[4]
w_5 <- w[5]
```

We can assign returns by pulling out columns from the `asset_returns_xts` object and then run the equation.

```
asset1 <- asset_returns_xts[,1]
asset2 <- asset_returns_xts[,2]
asset3 <- asset_returns_xts[,3]
asset4 <- asset_returns_xts[,4]
asset5 <- asset_returns_xts[,5]

portfolio_returns_byhand <-
  (w_1 * asset1) +
  (w_2 * asset2) +
  (w_3 * asset3) +
  (w_4 * asset4) +
  (w_5 * asset5)

names(portfolio_returns_byhand) <- "returns"
```

Our by-hand method is complete and we have monthly portfolio returns starting on January 31, 2013 through December 31, 2017. Now we can move on to the three universes.

3.1 Portfolio Returns in the xts world

For our first universe, we will use `Return.portfolio()` from `PerformanceAnalytics`, to calculate portfolio returns. The function requires two arguments for a portfolio, an xts object of returns and a vector of weights. It is not necessary but we are also going to set `rebalance_on = "months"` so we can confirm it matches our by-hand calculations above.

Remember, in the by-hand calculation, we set the portfolio weights as fixed, meaning they never changed on a month-to-month basis. That is equivalent to rebalancing every month. In practice, that would be quite rare. If we want a more realistic scenario, we could choose annual rebalancing by changing the argument to `rebalance_on = "years"`.

```
portfolio_returns_xts_rebalanced_monthly <-
  Return.portfolio(asset_returns_xts,
                   weights = w,
                   rebalance_on = "months") %>%
  `colnames<-`("returns")

head(portfolio_returns_xts_rebalanced_monthly, 3)
```

```
                returns
2013-01-31   0.0308488
2013-02-28  -0.0008697
2013-03-31   0.0186623
```

3.2 Portfolio Returns in the tidyverse

We begin our tidyverse work with our tidy data frame `asset_returns_long`. Our first task is to add a weights column to the `tibble` using the `mutate()` function. Each asset should be weighted according to the `w` vector. We use `case_when()` to assign weights by asset, so that in the case when the asset column is equal to asset 1 (SPY), we assign a weight of 0.25, or `w[1]`, and so on.

```
asset_returns_long %>%
group_by(asset) %>%
mutate(weights = case_when(asset == symbols[1] ~ w[1],
                           asset == symbols[2] ~ w[2],
                           asset == symbols[3] ~ w[3],
                           asset == symbols[4] ~ w[4],
                           asset == symbols[5] ~ w[5])) %>%
  head(3)
```

```
# A tibble: 3 x 4
# Groups:    asset [1]
  date         asset returns weights
  <date>       <chr>  <dbl>   <dbl>
1 2013-01-31 SPY     0.0499   0.250
2 2013-02-28 SPY     0.0127   0.250
3 2013-03-31 SPY     0.0373   0.250
```

Next, we need to implement the equation for portfolio returns. This task
is a bit tricky but serves as a nice way to use the group_by() function
with dates. We first add a new locum called weighted_returns that is the
product of each asset's monthly return and its weight. Then, we group_by()
the date column because each of our weighted returns needs to be added
together for each date. Once we group by date, we can use summarise(total
= sum(weighted_returns)) to add up the monthly weighted returns.

```
portfolio_returns_dplyr_byhand <-
  asset_returns_long %>%
  group_by(asset) %>%
  mutate(weights = case_when(asset == symbols[1] ~ w[1],
                             asset == symbols[2] ~ w[2],
                             asset == symbols[3] ~ w[3],
                             asset == symbols[4] ~ w[4],
                             asset == symbols[5] ~ w[5]),
             weighted_returns = returns * weights) %>%
  group_by(date) %>%
  summarise(returns = sum(weighted_returns))

head(portfolio_returns_dplyr_byhand, 3)
```

```
# A tibble: 3 x 2
  date         returns
  <date>        <dbl>
1 2013-01-31  0.0308
2 2013-02-28 -0.000870
```

```
3 2013-03-31   0.0187
```

That piped workflow required some logical hoops but it useful to see how to add those weights and then group by the date for finding total returns. Think about how we would solve the puzzle of rebalancing weights not every month, but every year.

3.3   Portfolio Returns in the tidyquant world

In tidyquant, we start again with our long, tidy-formatted `asset_returns_long` object, but convert to portfolio returns using `tq_portfolio()`.

The `tq_portfolio` function takes a `tibble` and then asks for an assets column to group by, a returns column to find return data, and a weights vector. It's a wrapper for `Return.portfolio()` and thus also accepts the argument `rebalance_on = "months"`. Since we are rebalancing by months, we should again get a portfolio returns object that matches our existing objects.

```
portfolio_returns_tq_rebalanced_monthly <-
  asset_returns_long %>%
  tq_portfolio(assets_col  = asset,
               returns_col = returns,
               weights     = w,
               col_rename  = "returns",
               rebalance_on = "months")
```

Let's take a quick look and compare how the tidy `tibbles` of portfolio returns compare to the `xts` objects of portfolio returns.

```
portfolio_returns_dplyr_byhand %>%
  rename(tidyverse = returns) %>%
  mutate(equation = coredata(portfolio_returns_byhand),
         tq = portfolio_returns_tq_rebalanced_monthly$returns,
         xts =
           coredata(portfolio_returns_xts_rebalanced_monthly)) %>%
  mutate_if(is.numeric, funs(round(., 3))) %>%
  head(3)
```

```
# A tibble: 3 x 5
  date       tidyverse equation      tq     xts
  <date>         <dbl>    <dbl>   <dbl>   <dbl>
```

```
1 2013-01-31    0.0310    0.0310    0.0310    0.0310
2 2013-02-28   -0.00100  -0.00100  -0.00100  -0.00100
3 2013-03-31    0.0190    0.0190    0.0190    0.0190
```

We have four objects of portfolio returns, calculated in four different ways, and with the same results.

3.4   Visualizing Portfolio Returns in the xts world

Let's start with `highcharter` to visualize the `xts` formatted portfolio returns.

First, we set `highchart(type = "stock")` to get a nice time series line. Then we add our **returns** column from `portfolio_returns_xts_rebalanced_monthly`. `highcharter` recognizes the date index so we do not need to point to it.

```
highchart(type = "stock") %>%
  hc_title(text = "Portfolio Monthly Returns") %>%
  hc_add_series(portfolio_returns_xts_rebalanced_monthly$returns,
                name = "Rebalanced Monthly",
                color = "cornflowerblue") %>%
  hc_add_theme(hc_theme_flat()) %>%
  hc_navigator(enabled = FALSE) %>%
  hc_scrollbar(enabled = FALSE) %>%
  hc_legend(enabled = TRUE) %>%
  hc_exporting(enabled = TRUE)
```

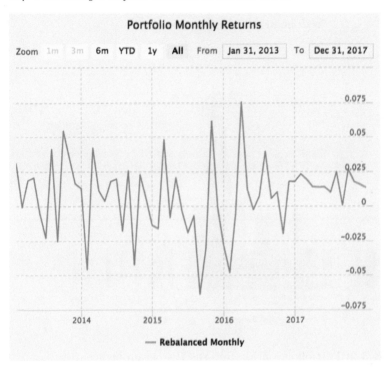

FIGURE 3.1: Portfolio Monthly Returns Line Chart highcharter

Figure 3.1 is a nice and simple line chart.

Let's use `highcharter` for histogram making, with the same code flow as we used previously.

```
hc_portfolio <-
  hist(portfolio_returns_xts_rebalanced_monthly$returns,
                 breaks = 50,
                 plot = FALSE)

hchart(hc_portfolio,
       color = "cornflowerblue",
       name = "Portfolio") %>%
  hc_title(text = "Portfolio Returns Distribution") %>%
  hc_add_theme(hc_theme_flat()) %>%
  hc_exporting(enabled = TRUE)
```

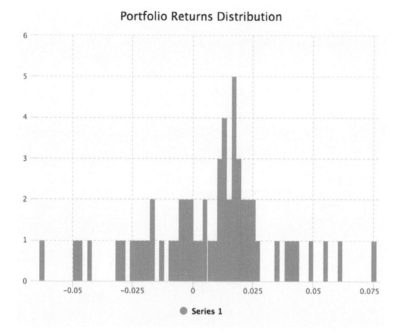

FIGURE 3.2: Portfolio Monthly Returns Histogram highcharter

As we see in Figure 3.2, `highcharter` does a nice job when there is one set of returns to place in a histogram.

3.5   Visualizing Portfolio Returns in the tidyverse

Let's start our `ggplot()` work with a scatter plot, a good way to investigate general trends in monthly returns. We need to supply both a x- and y-axis value for this chart.

```
portfolio_returns_tq_rebalanced_monthly %>%
  ggplot(aes(x = date, y = returns)) +
  geom_point(colour = "cornflowerblue")+
  xlab("date") +
  ylab("monthly return") +
  theme_update(plot.title = element_text(hjust = 0.5)) +
  ggtitle("Portfolio Returns Scatter") +
  scale_x_date(breaks = pretty_breaks(n=6))
```

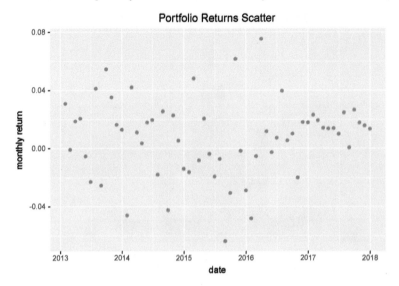

FIGURE 3.3: Portfolio Scatter Plot

The scatter plot in Figure 3.3 indicates that our portfolio did well in 2017, as we see zero negative returns for that year.

Let's move on to a histogram and use the same code flow as we did for the asset returns histogram.

```
portfolio_returns_tq_rebalanced_monthly %>%
  ggplot(aes(x = returns)) +
  geom_histogram(binwidth = .005,
                 fill = "cornflowerblue",
                 color = "cornflowerblue") +
  ggtitle("Portfolio Returns Distribution") +
  theme_update(plot.title = element_text(hjust = 0.5))
```

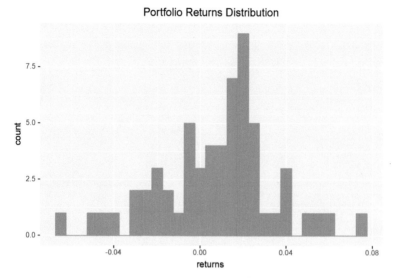

FIGURE 3.4: Portfolio Returns ggplot Histogram

Figure 3.4 shows that our most extreme monthly return was near .08.

We can also compare the portfolio distribution to those of our individual assets by layering on geoms. In figure 3.5, we use the `alpha` argument to make the asset histograms a bit faded, since there are more of them and the portfolio return is what we really want to see.

```
asset_returns_long %>%
  ggplot(aes(x = returns,
             fill = asset)) +
  geom_histogram(alpha = 0.15,
                 binwidth = .01) +
  geom_histogram(data = portfolio_returns_tq_rebalanced_monthly,
                 fill = "cornflowerblue",
                 binwidth = .01) +
  ggtitle("Portfolio and Asset Monthly Returns") +
  theme_update(plot.title = element_text(hjust = 0.5))
```

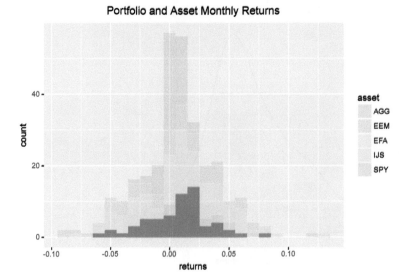

FIGURE 3.5: Asset and Portfolio Returns ggplot Histogram

As we did with individual returns, we can also put a histogram and density of portfolio returns on one chart.

We do that by first calling `geom_histogram()` then adding another layer with `geom_density()`.

```
portfolio_returns_tq_rebalanced_monthly %>%
  ggplot(aes(x = returns)) +
  geom_histogram(binwidth = .01,
                 colour = "cornflowerblue",
                 fill = "cornflowerblue") +
  geom_density(alpha = 1, color = "red") +
  xlab("monthly returns") +
  ylab("distribution") +
  theme_update(plot.title = element_text(hjust = 0.5)) +
  ggtitle("Portfolio Histogram and Density")
```

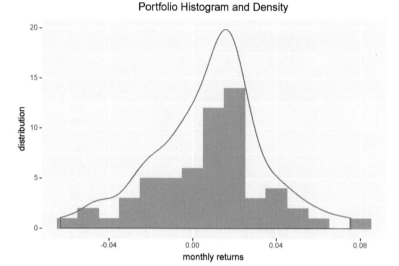

FIGURE 3.6: Portfolio Returns ggplot Histogram and Density

Figure 3.6 shows again how we can layer histogram and density geoms onto one chart.

That concludes our visualizations of portfolio returns. The code flows should have looked very familiar from our asset visualization work. We have done some nice exploratory work on our portfolio and now head to Shiny so that an end user (or even ourselves) can test out these visualizations on custom portfolios.

3.6   Shiny App Portfolio Returns

A Shiny application is a flexible, useful and powerful way to share our work. It is an interactive web application, which means we are about to become web programmers. In this section, we will build a Shiny app to display portfolio returns based on user inputs. Since this is our first Shiny app, we will review the code in detail and then reuse much of this code in future apps where we want to display different visualizations and statistics.

We want to empower an end user to do the following:

1)   build a portfolio by choosing assets and weights
2)   choose a start date

3) choose a rebalancing frequency
4) calculate portfolio returns
5) visualize the portfolio returns on histogram and density chart

The final app is shown in Figure 3.7.

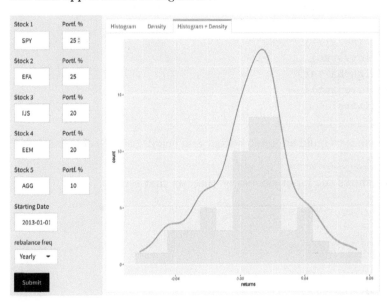

FIGURE 3.7: www.reproduciblefinance.com/shiny/returns

This encompasses much of our work thus far as it requires importing daily price data, converting to monthly log returns, assigning portfolio weights, calculating portfolio returns, and visualizing with `ggplot()`. The user can choose any 5 assets and our app could easily support 50 assets, though consider the user experience there - will any user manually enter 50 ticker symbols? At that number of assets, the preference would probably be to upload a csv file.

We will use RMarkdown to build our Shiny applications by inserting into the yaml `runtime: shiny`. This will alert the server (or our laptop) that this is an interactive document. The yaml also gives us a space for the title and to specify the format as `flexdashboard`. This is what the yaml looks like for the app (and the yaml for all of our future apps will be identical, except for the title):

```
---
title: "Returns Shiny"
runtime: shiny
output:
  flexdashboard::flex_dashboard:
```

```
    orientation: rows
---
```

We start with a **setup** code chunk to load packages. All of our Shiny apps will contain this code chunk because we need to load packages there.

```
# This is the setup chunk
library(tidyverse)
library(highcharter)
library(tidyquant)
library(timetk)
```

Our first task is to build an input sidebar and enable users to choose five stocks and weights.

Figure 3.8 shows the input sidebar as the user first sees it.

FIGURE 3.8: Input Sidebar from Shiny app

The code for building that sidebar starts with `textInput("stock1",...))` to create a space where the user can type a stock symbol and then `numericInput("w1",...)` to create a space where the user can enter a numeric weight. We want those entry spaces to be on the same line so we will nest them inside of a call to `fluidRow()`.

Since we have 5 stocks and weights, we repeat this 5 times. Notice that the stock symbol field uses `textInput()` because the user needs to enter text and the weight field uses `numericInput()` because the user needs to enter a number.

```
fluidRow(
  column(6,
  textInput("stock1", "Stock 1", "SPY")),
  column(5,
  numericInput("w1", "Portf. %", 25,
              min = 1, max = 100))
)

# Repeat this fluidRow() four more times, changing names to
# stock2, stock3, stock4, stock5 and w2, w3, 4, w5
```

Let's dissect one of those fluid rows line-by-line.

`fluidRow()` creates the row.

`column(6...)` creates a column for our stock ticker input with a length of 6.

`textInput("stock1", "Stock 1", "SPY"))` creates our first text input field.

We assigned it `stock1` which means it will be referenced in downstream code as `input$stock1`. We labeled it with "Stock 1", which is what the end user will see when viewing the app.

Finally, we set "SPY" as the default initial value. If the user does nothing, the value will be this default.

We also include a row where the user can choose a start date with `dateInput(...)`.

```
fluidRow(
  column(7,
  dateInput("date",
            "Starting Date",
            "2013-01-01",
            format = "yyyy-mm-dd"))
)
```

We give the user the ability to rebalance the portfolio at different intervals with `selectInput("rebalance", "rebal freq", c("Yearly" = "years", "Monthly" = "months"...)` and end with a submit button.

```
fluidRow(
  column(6,
  selectInput("rebalance", "rebal freq",
              c("Yearly" = "years",
                "Monthly" = "months",
```

```
                    "Weekly" = "weeks"))
  )
)
```

```
actionButton("go", "Submit")
```

The 'submit' button is very important because it enables the use of
`eventReactive()` to control our computation. An `eventReactive()` is a reactive function that will not start until it observes some event. In the next code
chunk, we tell `portfolio_returns_byhand` to wait for input$go by calling
`eventReactive(input$go....` Now, have a quick look back at the previous
code chunk, and note that we have `actionButton("go"...)`. Our reactive is
waiting for the user to click on the submit button we have labeled with **go**.

After that, the code chunk below should look very familiar from our previous
work, except it depends on user inputs for symbols, weights and starting date.

For example, when we previously built our portfolio, we statically defined
symbols as `symbols <- c("SPY", "EFA", "IJS", "EEM", "AGG")`.

In the chunk below, it is defined reactively as `symbols <- c(input$stock1,
input$stock2, input$stock3, input$stock4, input$stock5)`.

I copy the full code below even though it is very similar to how we calculated
portfolio returns in the non-Shiny context. For future Shiny apps, we will not
be reviewing this code again but they will all use a similar flow to take tickers
and weights for constructing a portfolio. Take a close look and identify how our
tickers, weights and starting date get passed to the `eventReactive()` function.
The tickers are input$stock1, input$stock2, etc, the weights are input$w1,
input$w2, etc. The date is input$date.

```
portfolio_returns_byhand <- eventReactive(input$go, {

  symbols <- c(input$stock1, input$stock2,
               input$stock3, input$stock4,
               input$stock5)

  prices <- getSymbols(symbols,
                       src = 'yahoo',
                       from = input$date,
                       auto.assign = TRUE,
                       warnings = FALSE) %>%
  map(~Ad(get(.))) %>%
  reduce(merge) %>%
    `colnames<-`(symbols)
```

```
w <- c(input$w1/100, input$w2/100,
       input$w3/100, input$w4/100,
       input$w5/100)

asset_returns_long <-
    prices %>%
    to.monthly(indexAt = "last",
               OHLC = FALSE) %>%
    tk_tbl(preserve_index = TRUE,
           rename_index = "date") %>%
    gather(asset, returns, -date) %>%
    group_by(asset) %>%
    mutate(returns = (log(returns) - log(lag(returns))))

portfolio_returns_byhand <-
    asset_returns_long %>%
    tq_portfolio(assets_col = asset,
                 returns_col = returns,
                 weights = w,
                 col_rename = "returns")

})
```

We now have a reactive called `portfolio_returns_byhand()` and we can pass that to our downstream code chunks to display distributions of portfolio returns. Thus far, our substantive work is very similar to what we did in the non-Shiny context throughout this section, except it depends on inputs from the user.

Next, we build charts to display our results interactively. Since we are not building any line charts, we will use `ggplot()` for this app and wait to use `highcharter` in future apps.

Shiny uses a custom function for building reactive `ggplots` called `renderPlot()`. By including `renderPlot()` in the code chunks, we are alerting the app that a reactive plot is being built, one that will change when an upstream reactive or input changes. In this case, the plot will change when the user clicks 'submit' and fires off the `eventReactive()`.

The flow for the three `ggplot()` code chunks, which appear in the different tabs, is going to be the same: call the reactive function `renderPlot()`, pass in `portfolio_returns_byhand()`, call `ggplot()` with an `aes(x = ...)` argument and then choose the appropriate `geom_`. The specifics of the `geom_` and other aesthetics are taken straight from our previous visualizations.

Here is the histogram code chunk.

```
renderPlot({
  portfolio_returns_byhand() %>%
    ggplot(aes(x = returns)) +
    geom_histogram(alpha = 0.25,
                   binwidth = .01,
                   fill = "cornflowerblue")
})
```

Here is the density chart code chunk.

```
renderPlot({
  portfolio_returns_byhand() %>%
    ggplot(aes(x = returns)) +
    geom_density(
                 size = 1,
                 color = "cornflowerblue")
})
```

Finally, here is the histogram and density on the same chart.

```
renderPlot({
  portfolio_returns_byhand() %>%
    ggplot(aes(x = returns)) +
    geom_histogram(alpha = 0.25,
                   binwidth = .01,
                   fill = "cornflowerblue") +
    geom_density(
                 size = 1,
                 color = "red")
})
```

Note how those previous 3 code chunks are the same as the visualization work we did in the non-Shiny context, except we wrap renderPlot() around the code flow and pass in the reactive portfolio_returns_byhand() instead of a tibble. That was by design: when building static visualizations as part of exploratory or testing phases, we always have Shiny in the back of our minds and try to use flows that will facilitate user inputs.

This Shiny app allows an end user to apply all of our work to a custom portfolio. It is a very simple app because we just ported our monthly returns to charts but it lays the foundation for more complex apps. The basic flow will not change: take user inputs, build a portfolio, run the portfolio through functions, visualize the results.

Concluding Returns

We have reviewed several paths, packages and code flows for building a multi-asset portfolio and calculating monthly log returns. At this point, you should feel comfortable with the difference between an `xts` object and a `tibble`, how to import prices, transform to returns, and employ various visualization techniques.

From a general data science paradigm perspective, we can think of this as data import, wrangling and transformation where:

(i)   pulling daily prices from Yahoo! Finance, csv or xls = data import

(ii) isolating adjusted prices and converting to monthly prices = data wrangling

(iii) converting to log returns, portfolio returns = data transformation

We were painstaking about our process to provide ourselves and collaborators with a clear data provenance, plus a variety of code paths for visualizing and inspecting data.

In the following sections, we will see how having several base portfolio returns objects facilitates our more analytic work. Make sure the `tibble` and `xts` objects are familiar and intuitive because we will use them throughout the rest of the book without reviewing their lineage.

If you are firing up a new R session and want to run the code to build all of our base portfolio returns objects, you can grab the code, with no text or explanations, here:

www.reproduciblefinance.com/code/get-returns/

Risk

Welcome to our section on risk, in which we calculate the standard deviation, skewness and kurtosis of portfolio returns. We will spend most of our time on standard deviation because it functions as the main measure of portfolio risk. To learn more about why, head back to 1959 and read Markowitz's monograph *Portfolio Selection: Efficient Diversification of Investments*,[1] which talks about means and variances of returns. In short, standard deviation measures the extent to which a portfolio's returns are dispersed around their mean. If returns are more dispersed, the portfolio has a higher standard deviation and is seen as riskier or more volatile.[2]

We will also discuss skewness and kurtosis, two indicators of how a portfolio's returns are distributed. Both add to the risk story and let us know about the dispersion of our returns. The code flows for skewness and kurtosis are similar to that for standard deviation, which means we can be more efficient and less wordy in those chapters.

From a toolkit perspective, we will accomplish the following in this section:

1) calculate and visualize standard deviation and rolling standard deviation
2) calculate and visualize skewness and rolling skewness
3) calculate and visualize kurtosis and rolling kurtosis
4) build Shiny apps for standard deviation, skewness and kurtosis

To analogize this section onto a general data science workflow, where the previous section mapped to data import and wrangling, this section maps to exploring and visualizing descriptive statistics of our portfolio. Standard deviation, skewness and kurtosis are descriptive statistics with respect to the variability of portfolio returns, and variability translates to risk for a portfolio.

We will be working with the portfolio returns objects that were created in the previous section. If you are firing up a new R session and want to run the code to build those objects, you can grab the code, with no explanation, here:

www.reproduciblefinance.com/code/get-returns/

---

[1] Markowitz, Harry. Portfolio Selection: Efficient Diversification of Investments, John Wiley & Sons, 1959.

[2] For more on volatility, see here: ftalphaville.ft.com/2018/02/28/1519839805000/An-abridged–illustrated-history-of-volatility/

4

▉▉▉▉▉▉▉▉▉▉▉▉▉▉▉▉▉▉▉▉

Standard Deviation

We start with the textbook equation for the standard deviation of a multi-asset portfolio.

$$Standard\ Deviation = \sqrt{\sum_{t=1}^{n}(x_i - \overline{x})^2/n}$$

where x is each monthly portfolio return, x-bar is the mean monthly portfolio return and n is the number of observations. For a multi-asset portfolio, that equation equates to the weight squared of each asset multiplied by that asset's variance, plus the covariance of each asset pair times the weight of each asset in the pair. It is a bit tedious to write that by hand with R code so we will not cover it here,[1] but we will walk through a matrix algebra code flow.

First, we build a covariance matrix of returns using the cov() function.

```
covariance_matrix <- cov(asset_returns_xts)
round(covariance_matrix, 5)
```

```
            SPY       EFA      IJS       EEM      AGG
SPY    0.00074  0.00070  0.00083  0.00068  -0.00001
EFA    0.00070  0.00106  0.00065  0.00104   0.00004
IJS    0.00083  0.00065  0.00157  0.00064  -0.00008
EEM    0.00068  0.00104  0.00064  0.00176   0.00011
AGG   -0.00001  0.00004 -0.00008  0.00011   0.00007
```

AGG, the US bond ETF, has a negative or very low covariance with the other ETFs and it should make a nice volatility dampener.

We now take the square root of the transpose of the weights vector times the covariance matrix times the weights vector. To perform matrix multiplication, we use %*%.

---

[1]For the curious, see here: www.reproduciblefinance.com/code/standard-deviation-by-hand/

```
sd_matrix_algebra <- sqrt(t(w) %*% covariance_matrix %*% w)

sd_matrix_algebra_percent <-
  round(sd_matrix_algebra * 100, 2) %>%
  `colnames<-`("standard deviation")

sd_matrix_algebra_percent[1,]
```

```
standard deviation
             2.66
```

It was not necessary but that matrix algebra code gives us a chance to look at the covariance matrix and consider which assets might help lower portfolio risk. We will return to matrix algebra in the final section of the book.

4.1   Standard Deviation in the xts world

In the xts paradigm, we can use the built-in StdDev() function from PerformanceAnalytics to go straight from asset returns to portfolio standard deviation. It takes two arguments, a vector of returns and weights: StdDev(asset_returns_xts, weights = w).

```
portfolio_sd_xts_builtin <-
  StdDev(asset_returns_xts, weights = w)

portfolio_sd_xts_builtin_percent <-
  round(portfolio_sd_xts_builtin * 100, 2)

portfolio_sd_xts_builtin_percent[1,1]
```

```
[1] 2.66
```

4.2   Standard Deviation in the tidyverse

For the tidyverse, we start with portfolio_returns_dplyr_byhand and use the summarise() and sd() functions. We will also perform the calculation with our own equation `sqrt(sum((returns -` `mean(returns))^2)/(nrow(.)-1)))` as an example of how to put a custom equation into the `summarise()` code flow.

```
portfolio_sd_tidy_builtin_percent <-
  portfolio_returns_dplyr_byhand %>%
  summarise(
    sd = sd(returns),
    sd_byhand =
      sqrt(sum((returns - mean(returns))^2)/(nrow(.)-1))) %>%
  mutate(dplyr = round(sd, 4) * 100,
         dplyr_byhand = round(sd_byhand, 4) * 100)

portfolio_sd_tidy_builtin_percent %>%
  select(dplyr, dplyr_byhand)
```

```
# A tibble: 1 x 2
  dplyr dplyr_byhand
  <dbl>        <dbl>
1  2.66         2.66
```

Two more calculations complete.

4.3   Standard Deviation in the tidyquant world

In the `tidyquant` flow, we start with `portfolio_returns_tq_rebalanced_monthly` and invoke the `table.Stats()` function from `PerformanceAnalytics` by way of `tq_performance()`. The `table.Stats()` function returns a table of statistics for the portfolio but since we want only standard deviation, we will use `dplyr` to `select()` the `Stdev` column.

```
portfolio_sd_tidyquant_builtin_percent <-
portfolio_returns_tq_rebalanced_monthly %>%
  tq_performance(Ra = returns,
                 Rb = NULL,
                 performance_fun = table.Stats) %>%
  select(Stdev) %>%
  mutate(tq_sd = round(Stdev, 4) * 100)
```

This nicely demonstrates how `tidyquant` blends together the xts and tidy paradigms. Here we can use a function from `PerformanceAnalytics` and apply it to a `tibble` of returns.

Let's review our calculations.

```
portfolio_sd_tidy_builtin_percent %>%
  select(dplyr, dplyr_byhand) %>%
  mutate(xts_builtin = portfolio_sd_xts_builtin_percent,
         matrix = sd_matrix_algebra_percent,
         tq = portfolio_sd_tidyquant_builtin_percent$tq_sd)
```

```
# A tibble: 1 x 5
  dplyr dplyr_byhand xts_builtin matrix    tq
  <dbl>        <dbl>       <dbl> <dbl> <dbl>
1  2.66         2.66        2.66  2.66  2.66
```

We should now feel comfortable with calculating portfolio standard deviation starting from different object types and using many code flows. Let's see how this work can be visualized.

4.4   Visualizing Standard Deviation

Visualizing standard deviation of portfolio returns comes down to visualizing the dispersion of portfolio returns.

Figure 4.1 shows the scatter plot of monthly returns that we built in the *Returns* section.

```
portfolio_returns_dplyr_byhand %>%
  ggplot(aes(x = date, y = returns)) +
  geom_point(color = "cornflowerblue") +
  scale_x_date(breaks = pretty_breaks(n = 6)) +
  ggtitle("Scatterplot of Returns by Date") +
  theme(plot.title = element_text(hjust = 0.5))
```

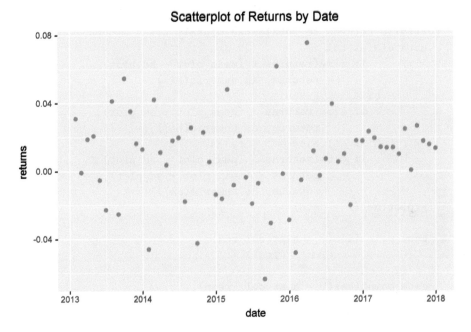

FIGURE 4.1: Dispersion of Portfolio Returns

At first glance and using an unscientific eyeball test, it seems that 2015 was the riskiest year. Let's add a different color for any monthly returns that are one standard deviation away from the mean.

First, we will create an indicator for the mean return with `mean()` and one for the standard deviation with `sd()`. We will call the variables `mean_plot` and `sd_plot` since we plan to use them in a plot and not for anything else.

```
sd_plot <-
  sd(portfolio_returns_tq_rebalanced_monthly$returns)
mean_plot <-
  mean(portfolio_returns_tq_rebalanced_monthly$returns)
```

We want to shade the scatter points according to a returns distance from the mean. Accordingly, we `mutate()` or create three new columns based on if-else logic. If the return is one standard deviation below the mean, we want to add that observation to the column we call `hist_col_red`, else that column should have an NA. We will create three new columns this way with `if_else()`.

Figure 4.2 shows the results when we color by these columns.

```
portfolio_returns_tq_rebalanced_monthly %>%
  mutate(hist_col_red =
           if_else(returns < (mean_plot - sd_plot),
                   returns, as.numeric(NA)),
         hist_col_green =
           if_else(returns > (mean_plot + sd_plot),
                   returns, as.numeric(NA)),
         hist_col_blue =
           if_else(returns > (mean_plot - sd_plot) &
                   returns < (mean_plot + sd_plot),
                   returns, as.numeric(NA))) %>%
  ggplot(aes(x = date)) +

  geom_point(aes(y = hist_col_red),
             color = "red") +

  geom_point(aes(y = hist_col_green),
             color = "green") +

  geom_point(aes(y = hist_col_blue),
             color = "blue") +
  labs(title = "Colored Scatter", y = "monthly returns") +
  scale_x_date(breaks = pretty_breaks(n = 8)) +
  theme(plot.title = element_text(hjust = 0.5))
```

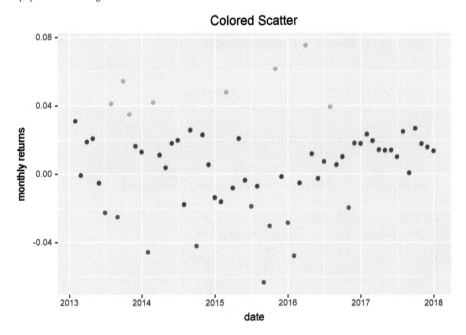

FIGURE 4.2: Scatter of Returns Colored by Distance from Mean

Figure 4.2 gives a little more intuition about the dispersion as we can see how many red and green dots appear.

Let's add a line for the value that is one standard deviation above and below the mean with `geom_hline(yintercept = (mean_plot + sd_plot), color = "purple"...)` + `geom_hline(yintercept = (mean_plot - sd_plot), color = "purple"...)` +.

```
portfolio_returns_tq_rebalanced_monthly %>%
    mutate(hist_col_red =
            if_else(returns < (mean_plot - sd_plot),
                    returns, as.numeric(NA)),
          hist_col_green =
            if_else(returns > (mean_plot + sd_plot),
                    returns, as.numeric(NA)),
          hist_col_blue =
            if_else(returns > (mean_plot - sd_plot) &
                    returns < (mean_plot + sd_plot),
                    returns, as.numeric(NA))) %>%

    ggplot(aes(x = date)) +
```

```
geom_point(aes(y = hist_col_red),
           color = "red") +

geom_point(aes(y = hist_col_green),
           color = "green") +

geom_point(aes(y = hist_col_blue),
           color = "blue") +

geom_hline(yintercept = (mean_plot + sd_plot),
           color = "purple",
           linetype = "dotted") +
geom_hline(yintercept = (mean_plot-sd_plot),
           color = "purple",
           linetype = "dotted") +
labs(title = "Colored Scatter with Line", y = "monthly returns") +
scale_x_date(breaks = pretty_breaks(n = 8)) +
theme(plot.title = element_text(hjust = 0.5))
```

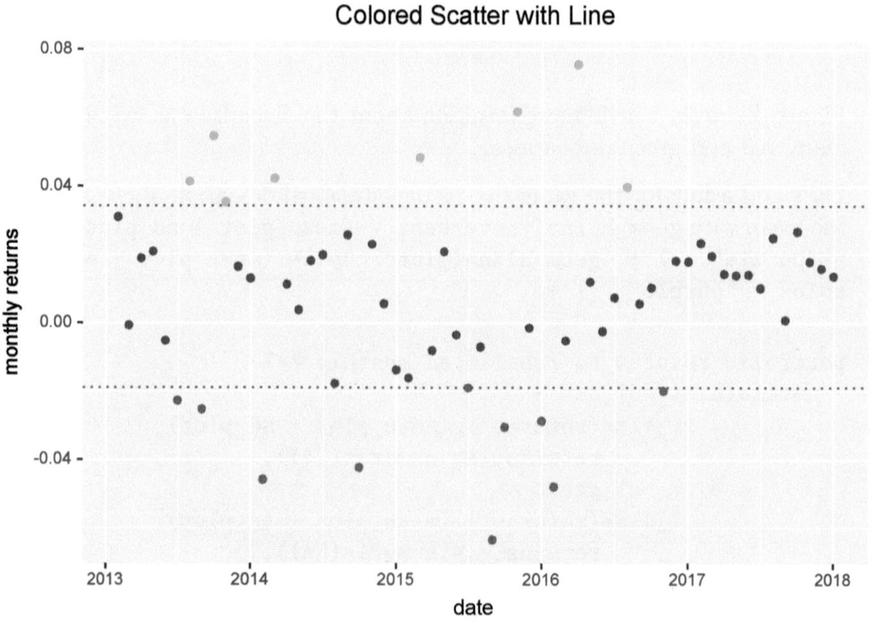

FIGURE 4.3: Scatter of Returns with Line at Standard Deviation

Figure 4.3 is showing us returns over time and whether they fall below or above one standard deviation from the mean. One element that jumps out is how many red or green circles we see after January 1, 2017. Zero! That is zero monthly returns that are least one standard deviation from the mean during calendar year 2017. When we get to rolling volatility, we should see this reflected as a low rolling volatility through 2017, along with high rolling volatility through 2015.

To visualize the actual standard deviation of our portfolio, it helps to do so in a comparative manner. In this case, we can explore how our portfolio risk compares to the risk of the 5 individual assets. First, we return to our `asset_returns_long` data frame and calculate the standard deviation of each asset's returns with `summarise(sd = sd(returns))`. Then we use `dplyr`'s `add_row()` to add the portfolio standard deviation from `portfolio_sd_tidy_builtin_percent` and end with a call to `ggplot()`.

```
asset_returns_long %>%
  group_by(asset) %>%
  summarize(sd = 100 *sd(returns)) %>%
  add_row(asset = "Portfolio",
          sd = portfolio_sd_tidy_builtin_percent$dplyr) %>%
  ggplot(aes(x = asset,
             y = sd,
             colour = asset)) +
  geom_point() +
  scale_y_continuous(labels = function(x) paste0(x, "%")) +
  geom_text(
        aes(x = "Portfolio",
            y =
               portfolio_sd_tidy_builtin_percent$dplyr + .2),
            label = "Portfolio",
          color = "cornflowerblue") +
  labs(y = "standard deviation")
```

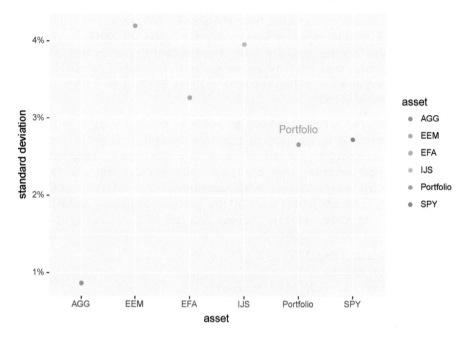

FIGURE 4.4: Asset and Portfolio Standard Deviation Comparison

Figure 4.4 is a nice example of when portfolio statistics are better viewed in comparison rather than in isolation.

We can also incorporate our work on returns by visualizing expected monthly returns (which is the mean return) scattered against standard deviation of monthly returns.

```
asset_returns_long %>%
  group_by(asset) %>%
  summarise(expected_return = mean(returns),
            stand_dev = sd(returns)) %>%
  add_row(asset = "Portfolio",
    stand_dev =
      sd(portfolio_returns_tq_rebalanced_monthly$returns),
    expected_return =
      mean(portfolio_returns_tq_rebalanced_monthly$returns)) %>%

ggplot(aes(x = stand_dev,
           y = expected_return,
           color = asset)) +
  geom_point(size = 2) +
```

```
geom_text(
  aes(x =
  sd(portfolio_returns_tq_rebalanced_monthly$returns) * 1.11,
    y =
  mean(portfolio_returns_tq_rebalanced_monthly$returns),
      label = "Portfolio")) +
ylab("expected return") +
xlab("standard deviation") +
ggtitle("Expected Monthly Returns versus Risk") +
scale_y_continuous(labels = function(x){ paste0(x, "%")}) +
# The next line centers the title
theme_update(plot.title = element_text(hjust = 0.5))
```

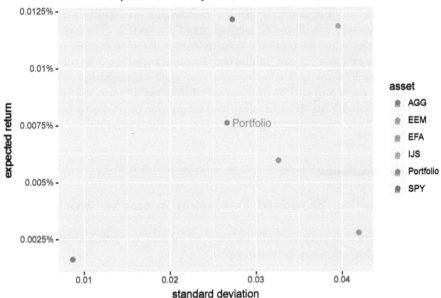

FIGURE 4.5: Expected Returns versus Risk

Figure 4.5 offers a nice look at our portfolio return versus risk profile (and it looks pretty good), though the S&P500 has a higher expected return for just a bit more risk. EEM and EFA have a higher risk and lower expected return (no rational investor wants that!) and IJS has a higher risk and a higher expected return (some rational investors do want that!).

4.5   Rolling Standard Deviation

We have calculated the average volatility for the entire life of the portfolio but
it would help if we could better understand how that volatility has changed
over time or behaved in different market conditions.

We might miss a 3-month or 6-month period where the volatility spiked or
plummeted or did both. And the longer our portfolio life, the more likely we
are to miss something important. If we had 10 or 20 years of data and we
calculated the standard deviation for the entire history, we could, or most
certainly would, fail to notice a period in which volatility was very high, and
hence we would fail to ponder the probability that it could occur again.

Imagine a portfolio which had a standard deviation of returns for each 6-month
period of 3% and it never changed. Now imagine a portfolio whose volatility
fluctuated every few 6-month periods from 0% to 6% . We might find a 3%
standard deviation of monthly returns over a 10-year sample for both, but
those two portfolios are not exhibiting the same volatility. The rolling volatility
of each would show us the differences and then we could hypothesize about
the past causes and future probabilities for those differences. We might also
want to think about dynamically rebalancing our portfolio to better manage
volatility if we are seeing large spikes in the rolling windows.

4.6   Rolling Standard Deviation in the xts world

The xts world is purpose-built for time series and, as such, calculating rolling
standard deviation is straightforward.

First, we assign a value of 24 to the variable window.

```
window <- 24
```

We then invoke rollapply(), pass it our xts returns object, the sd() function,
and a rolling window with width = window.

```
port_rolling_sd_xts <-
  rollapply(portfolio_returns_xts_rebalanced_monthly,
            FUN = sd,
            width = window) %>%
  # omit the 23 months for which there is no rolling 24
```

```
# month standard deviation
na.omit() %>%
`colnames<-`("rolling_sd")

tail(port_rolling_sd_xts, 3)
```

```
           rolling_sd
2017-10-31    0.02339
2017-11-30    0.02328
2017-12-31    0.02169
```

4.7 Rolling Standard Deviation in the tidyverse

In the tidyverse, rolling calculations with time series are difficult. Try the code below.

```
port_rolling_sd_tidy_does_not_work <-
  portfolio_returns_dplyr_byhand %>%
  mutate(rolling_sd = rollapply(returns,
                                FUN = sd,
                                width = window,
                                fill = NA)) %>%
  select(date, rolling_sd) %>%
  na.omit()

tail(port_rolling_sd_tidy_does_not_work, 3)
```

```
# A tibble: 3 x 2
   date        rolling_sd
   <date>          <dbl>
1 2016-10-31      0.0234
2 2016-11-30      0.0233
3 2016-12-31      0.0217
```

That does *not* match our xts result.

The width argument is not picked up correctly and the year 2017 disappeared. Tibbles were not built for time series analysis.

4.8 Rolling Standard Deviation with the tidyverse and tibbletime

Tibbletime and its `rollify()` function *are* built for time series analysis and allow us to solve this problem.

We use `rollify()` to define a rolling standard deviation function. We want to roll the `sd()` function with a width equal to `window` so we define `sd_roll_24 <- rollify(sd, window = window)`.

```
sd_roll_24 <-
  rollify(sd, window = window)
```

Then we use `mutate()` to pass it into the code flow. Note that we convert our tibble to a tibbletime data frame with `as_tbl_time(index = date)`.

```
port_rolling_sd_tidy_tibbletime <-
  portfolio_returns_tq_rebalanced_monthly %>%
  as_tbl_time(index = date) %>%
  mutate(sd = sd_roll_24(returns)) %>%
  select(-returns) %>%
  na.omit()

tail(port_rolling_sd_tidy_tibbletime, 3)

# A time tibble: 3 x 2
# Index: date
  date            sd
  <date>       <dbl>
1 2017-10-31 0.0234
2 2017-11-30 0.0233
3 2017-12-31 0.0217
```

That nifty combination of the tidyverse and `tibbletime` is generalizable to other functions beyond standard deviation. Tibbletime is changing and improving rapidly as of the time of this writing (Spring of 2018). We will keep an eye on the package and post new use cases to the website as things develop. Stay tuned!

4.9   Rolling Standard Deviation in the tidyquant world

The tidyquant package has a nice way to apply a rolling function to data frames as well. We take tq_mutate() and supply mutate_fun = rollapply as our mutation function argument. Then, we invoke FUN = sd as the nested function beneath rollapply().

```
port_rolling_sd_tq <-
  portfolio_returns_tq_rebalanced_monthly %>%
  tq_mutate(mutate_fun = rollapply,
            width = window,
            FUN = sd,
            col_rename = "rolling_sd") %>%
  select(date, rolling_sd) %>%
  na.omit()
```

Take a quick peek to confirm consistent results.

```
port_rolling_sd_tidy_tibbletime %>%
  mutate(sd_tq = port_rolling_sd_tq$rolling_sd,
         sd_xts = round(port_rolling_sd_xts$rolling_sd, 4)) %>%
  tail(3)
```

```
# A time tibble: 3 x 4
# Index: date
  date            sd    sd_tq sd_xts
  <date>       <dbl>    <dbl> <S3: xts>
1 2017-10-31 0.0234 0.0234 0.0234
2 2017-11-30 0.0233 0.0233 0.0233
3 2017-12-31 0.0217 0.0217 0.0217
```

We now have an xts object called port_rolling_sd_xts, a tibbletime tibble called port_rolling_sd_tidy_tibbletime and a tibble object called port_rolling_sd_tq. Each contains the 24-month rolling standard deviation of portfolio returns.

At the outset of this section, we opined that rolling volatility might add some insight that is obscured by the total volatility. Visualizing the rolling standard deviation should help to illuminate this and that is where we head next.

4.10 Visualizing Rolling Standard Deviation in the xts world

We begin our visualization work with `highcharter`.

First, we convert to our data to rounded percentages for ease of charting.

```
port_rolling_sd_xts_hc <-
  round(port_rolling_sd_xts, 4) * 100
```

Then we return to the familiar invocation of the `highcharter` flow.

```
highchart(type = "stock") %>%
  hc_title(text = "24-Month Rolling Volatility") %>%
  hc_add_series(port_rolling_sd_xts_hc,
                 color = "cornflowerblue") %>%
  hc_add_theme(hc_theme_flat()) %>%
  hc_yAxis(
    labels = list(format = "{value}%"),
             opposite = FALSE) %>%
  hc_navigator(enabled = FALSE) %>%
  hc_scrollbar(enabled = FALSE) %>%
  hc_exporting(enabled= TRUE) %>%
  hc_legend(enabled = TRUE)
```

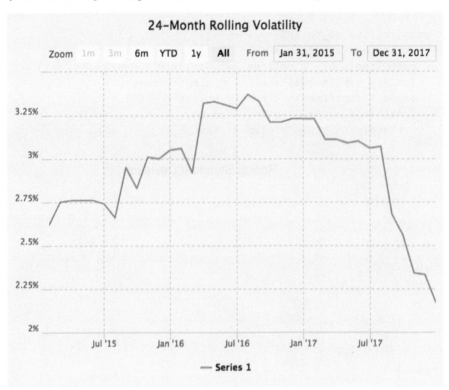

FIGURE 4.6: Rolling Portfolio Volatility highcharter

Figure 4.6 visualization offers a nice history of the 24-month portfolio volatility. We see it rising through 2015 and early 2016, then plummeting through 2017.

4.11 Visualizing Rolling Standard Deviation in the tidyverse

We begin our tidyverse visualization by passing `port_rolling_sd_tq` to `ggplot()`. We want to chart rolling standard deviation as a line chart, with date on the x-axis. We call `aes(x=date)` and then `geom_line(aes(y = rolling_sd), color = "cornflowerblue")`.

```
port_rolling_sd_tq %>%
  ggplot(aes(x = date)) +
  geom_line(aes(y = rolling_sd), color = "cornflowerblue") +
  scale_y_continuous(labels = scales::percent) +
  scale_x_date(breaks = pretty_breaks(n = 8)) +
  labs(title = "Rolling Standard Deviation", y = "") +
    theme(plot.title = element_text(hjust = 0.5))
```

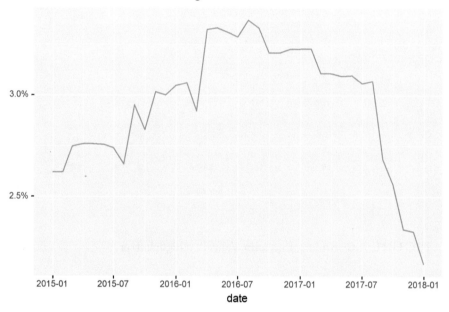

FIGURE 4.7: Rolling Volatility ggplot

Look at Figure 4.7 and note that we did not manually change our decimal to percentage format. When we called `scale_y_continuous(labels = scales::percent)` it did that work for us by adding the % sign and multiplying by 100.

Do these visualizations add to our understanding of this portfolio? Well, we can see a spike in rolling volatility in 2016 followed by a consistently falling volatility through mid-2017. That makes sense. Remember back to our returns dispersion scatter plot when zero monthly returns were more than one standard deviation away from the mean in 2017.

4.12 Shiny App Standard Deviation

Now let's wrap all of that work into a Shiny app that allows a user to choose a 5-asset portfolio and chart rolling volatility of different widths.

Have a look at the app in Figure 4.8:

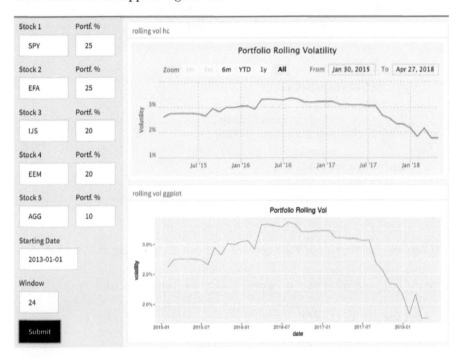

FIGURE 4.8: www.reproduciblefinance.com/shiny/standard-deviation/

Our input sidebar is almost identical to our previous app, with the addition that we let the user choose a rolling window with a `numericInput()`. The chunk below is rolling window snippet of our sidebar code. Notice I set `min = 3, max = 36` so the user must choose a rolling window between 3 and 36 months.

```
fluidRow(
  column(5,
  numericInput("window", "Window", 12,
              min = 3, max = 36, step = 1))
)
```

Next we calculate rolling volatility with the same work flow as above. Note we are using the `tidyquant` method and when we chart with `highcharter`, we will convert to `xts`.

First, we get stock prices.

```
prices <- eventReactive(input$go, {

  symbols <- c(input$stock1, input$stock2,
               input$stock3, input$stock4,
               input$stock5)

  prices <-
    getSymbols(symbols,
               src = 'yahoo',
               from = input$date,
               auto.assign = TRUE,
               warnings = FALSE) %>%
    map(~Ad(get(.))) %>%
    reduce(merge) %>%
    `colnames<-`(symbols)
})
```

Then we convert to portfolio returns and calculate rolling standard deviation.

```
port_rolling_sd_tidy <- eventReactive(input$go, {

  prices <- prices()

  w <- c(input$w1/100, input$w2/100,
         input$w3/100, input$w4/100,
         input$w5/100)

  portfolio_returns_tq_rebalanced_monthly <-
    prices %>%
    to.monthly(indexAt = "last",
               OHLC = FALSE) %>%
    tk_tbl(preserve_index = TRUE,
           rename_index = "date") %>%
    slice(-1) %>%
    gather(asset, returns, -date) %>%
    group_by(asset) %>%
    mutate(returns =
             (log(returns) - log(lag(returns)))) %>%
```

```
    tq_portfolio(assets_col  = asset,
                 returns_col = returns,
                 weights     = w,
                 col_rename  = "returns",
                 rebalance_on = "months")

  window <- input$window

  port_rolling_sd_tidy <-
  portfolio_returns_tq_rebalanced_monthly %>%
  tq_mutate(mutate_fun = rollapply,
            width = window,
            FUN = sd,
            col_rename = ("rolling_sd")) %>%
  select(date, rolling_sd) %>%
  na.omit()

})
```

We now have the object `port_rolling_sd_tidy()` available for downstream visualizations. We will start with `highcharter` and that requires changing to an `xts` object with `tk_xts()`.

Note that we start the code with `renderHighchart()`, instead of `renderPlot()`. That alerts Shiny that a reactive `highcharter` object will be displayed here.

```
renderHighchart({

  port_rolling_sd_xts_hc <-
    port_rolling_sd_tidy() %>%
    tk_xts(date_col = date) %>%
    round(., 4) * 100


  highchart(type = "stock") %>%
    hc_title(text = "Portfolio Rolling Volatility") %>%
    hc_yAxis(title = list(text = "Volatility"),
             labels = list(format = "{value}%"),
             opposite = FALSE) %>%
    hc_add_series(port_rolling_sd_xts_hc,
                  name = "Portfolio Vol",
                  color = "cornflowerblue",) %>%
    hc_add_theme(hc_theme_flat()) %>%
```

```
    hc_navigator(enabled = FALSE) %>%
    hc_scrollbar(enabled = FALSE) %>%
    hc_exporting(enabled = TRUE)
```

For our next visualization, we pass the `port_rolling_sd_tidy()` reactive straight into a `ggplot()` code flow.

```
renderPlot({
  port_rolling_sd_tidy() %>%
    ggplot(aes(x = date)) +
    geom_line(aes(y = rolling_sd),
              color = "cornflowerblue") +
    scale_y_continuous(labels =
                            scales::percent) +
    ggtitle("Portfolio Rolling Vol") +
    ylab("volatility") +
    scale_x_date(breaks =
                    pretty_breaks(n = 8)) +
    theme(plot.title =
            element_text(hjust = 0.5))
})
```

That completes our Shiny app on rolling portfolio volatility. We included two very similar plots in that app and probably would not do that in the real world, but here we can experiment and be over-inclusive with our aesthetics.

5

Skewness

Skewness is the degree to which returns are asymmetric around their mean. Since a normal distribution is symmetric around the mean, skewness can be taken as one measure of how returns are not distributed normally. Why does skewness matter? If portfolio returns are right, or positively, skewed, it implies numerous small negative returns and a few large positive returns. If portfolio returns are left, or negatively, skewed, it implies numerous small positive returns and few large negative returns.

Here's the equation for skewness:

$$Skew = \sum_{t=1}^{n} (x_i - \overline{x})^3 / n \Big/ (\sum_{t=1}^{n} (x_i - \overline{x})^2 / n)^{3/2}$$

Skewness has important substantive implications for risk and is also a concept that lends itself to data visualization. In fact, the visualizations are often more illuminating than the numbers themselves (though the numbers are what matter in the end). In this chapter, we will cover how to calculate skewness using xts and tidyverse methods, how to calculate rolling skewness and how to create several data visualizations as pedagogical aids.

5.1 Skewness in the xts world

We begin in the xts world and make use of the skewness() function from PerformanceAnalytics.

```
skew_xts <-
  skewness(portfolio_returns_xts_rebalanced_monthly$returns)

skew_xts
```

```
[1] -0.232
```

Our portfolio is relatively balanced and shows slight negative skewness, which does not seem cause for major concern. Let's head to the tidyverse and explore more code flows.

5.2   Skewness in the tidyverse

We will make use of the same `skewness()` function but because we start with a tibble, we use `summarise(skew = skewness(returns)`. We also run this calculation by-hand and save the result in a column called `skew_byhand`.

```
skew_tidy <-
  portfolio_returns_tq_rebalanced_monthly %>%
  summarise(skew_builtin = skewness(returns),
            skew_byhand =
(sum((returns - mean(returns))^3)/length(returns))/
((sum((returns - mean(returns))^2)/length(returns)))^(3/2)) %>%
  select(skew_builtin, skew_byhand)

skew_tidy %>%
  mutate(xts = coredata(skew_xts)) %>%
  mutate_all(funs(round(., 3)))

# A tibble: 1 x 3
  skew_builtin skew_byhand     xts
         <dbl>       <dbl>   <dbl>
1       -0.232      -0.232  -0.232
```

The results are consistent using `xts` and our tidy by-hand and built-in methods. As with `StdDev`, the `skewness()` function is not wrapped into `tidyquant` so we will omit that universe and head to visualization.

5.3   Visualizing Skewness

To visualize skewness, let's start with a histogram of returns and set `binwidth` = `.003` to get fine granularity.

```
portfolio_returns_tq_rebalanced_monthly %>%
ggplot(aes(x = returns)) +
geom_histogram(alpha = .7,
               binwidth = .003,
               fill = "cornflowerblue",
               color = "cornflowerblue") +
scale_x_continuous(breaks =
                   pretty_breaks(n = 10))
```

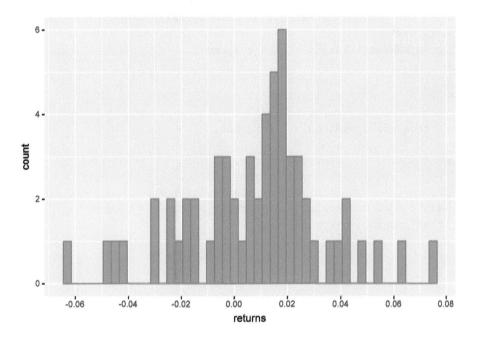

FIGURE 5.1: Returns Histogram

Figure 5.1 shows one highly negative return (worse than -.06) and one highly positive return (greater than .06), several between -/+.04 and -/+.06, plus a cluster of negative returns around -.02.

We can get more creative about which returns we want to highlight and investigate. For example, let's focus on monthly returns that fall 2 standard deviations below the mean by creating create three new columns: one for returns below a threshold, one for returns above a threshold, and one for returns within the two thresholds.

This code creates a column for each monthly return that is two standard deviations below the mean: `hist_col_red = if_else(returns <`

(mean(returns) - 2*sd(returns)), returns, NA). I labeled the new column hist_col_red because we will shade these red to connote very negative returns.

```
portfolio_returns_tq_rebalanced_monthly %>%
  mutate(hist_col_red =
        if_else(returns < (mean(returns) - 2*sd(returns)),
                returns, as.numeric(NA)),
         returns =
        if_else(returns > (mean(returns) - 2*sd(returns)),
                returns, as.numeric(NA))) %>%
ggplot() +
geom_histogram(aes(x = hist_col_red),
               alpha = .7,
               binwidth = .003,
               fill = "red",
               color = "red") +
  geom_histogram(aes(x = returns),
               alpha = .7,
               binwidth = .003,
               fill = "cornflowerblue",
               color = "cornflowerblue") +
scale_x_continuous(breaks = pretty_breaks(n = 10)) +
xlab("monthly returns")
```

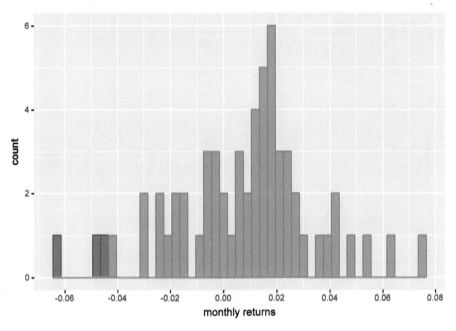

FIGURE 5.2: Shaded Histogram Returns

Figure 5.2 helps to identify what is driving the skewness but a density plot often accompanies skewness.

To build a density plot, we will call `stat_density()` in `ggplot()`.

```
portfolio_density_plot <-
  portfolio_returns_tq_rebalanced_monthly %>%
  ggplot(aes(x = returns)) +
  stat_density(geom = "line",
               alpha = 1,
               colour = "cornflowerblue")

portfolio_density_plot
```

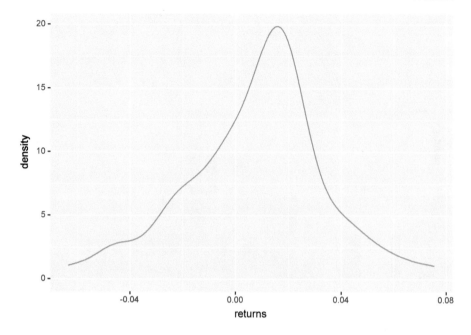

FIGURE 5.3: Density Plot Skewness

Figure 5.3 looks good as a simple density chart but it would be nice to shade the area that falls below some threshold again. To shade below the mean, we create an object called shaded_area using ggplot_build(portfolio_density_plot)$data[[1]] %>% filter(x < mean(portfolio_returns_tq_rebalanced_monthly$returns)). That snippet will take our original ggplot() object and create a new object filtered for x-values less than the mean return. Then we use geom_area to add the shaded area to portfolio_density_plot.

```
shaded_area_data <-
ggplot_build(portfolio_density_plot)$data[[1]] %>%
  filter(x <
       mean(portfolio_returns_tq_rebalanced_monthly$returns))

portfolio_density_plot_shaded <-
  portfolio_density_plot +
  geom_area(data = shaded_area_data,
            aes(x = x, y = y),
            fill="pink",
            alpha = 0.5)
```

```
portfolio_density_plot_shaded
```

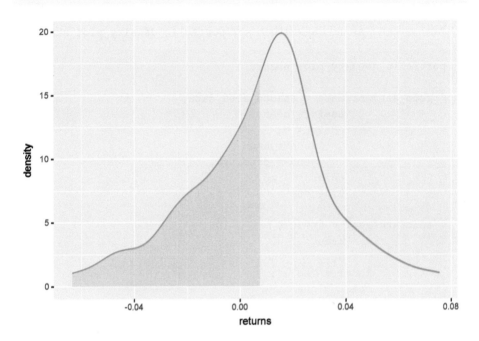

FIGURE 5.4: Density Plot with Shaded Area

The shaded area in Figure 5.4 highlights the mass of returns that fall below the mean.

We can add a vertical line at the mean and median, and some explanatory labels. This will help to emphasize that negative skew indicates a mean less than the median.

First, create variables for the mean and median.

```
median <-
  median(portfolio_returns_tq_rebalanced_monthly$returns)
mean <-
  mean(portfolio_returns_tq_rebalanced_monthly$returns)
```

We want the vertical lines to just touch the density plot.

```
median_line_data <-
```

```
ggplot_build(portfolio_density_plot)$data[[1]] %>%
filter(x <= median)
```

Now we can start adding aesthetics to `portfolio_density_plot_shaded`.

```
portfolio_density_plot_shaded +

  geom_segment(data = shaded_area_data,
               aes(x = mean,
                   y = 0,
                   xend = mean,
                   yend = density),
               color = "red",
               linetype = "dotted") +

  annotate(geom = "text",
           x = mean,
           y = 5,
           label = "mean",
           color = "red",
           fontface = "plain",
           angle = 90,
           alpha = .8,
           vjust =  -1.75) +

  geom_segment(data = median_line_data,
               aes(x = median,
                   y = 0,
                   xend = median,
                   yend = density),
               color = "black",
               linetype = "dotted") +

  annotate(geom = "text",
           x = median,
           y = 5,
           label = "median",
           fontface = "plain",
           angle = 90,
           alpha = .8,
           vjust =  1.75) +
  ggtitle("Density Plot Illustrating Skewness")
```

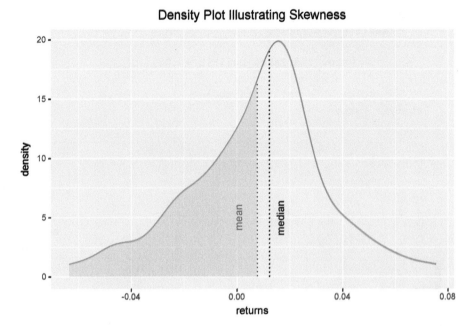

FIGURE 5.5: Density Plot Shaded with Lines

Figure 5.5 includes quite a bit of content, possibly too much, but better to be over-inclusive now to test different variants. We can delete any of those features when using this chart later or refer back to these lines of code should we ever want to reuse some of the aesthetics.

At this point, we have calculated the skewness using three methods and created some nice exploratory visualizations.

Now we can investigate how our portfolio's skewness compares to the 5 assets' skewness. The code flow is very similar to the comparison we ran for standard deviation, except we use the skewness function.

```
asset_returns_long %>%
  summarize(skew_assets = skewness(returns)) %>%
  add_row(asset = "Portfolio",
          skew_assets = skew_tidy$skew_byhand)%>%
  ggplot(aes(x = asset,
             y = skew_assets,
             colour = asset)) +
  geom_point() +
  geom_text(
```

```
        aes(x = "Portfolio",
            y =
            skew_tidy$skew_builtin + .04),
            label = "Portfolio",
          color = "cornflowerblue") +
  labs(y = "skewness")
```

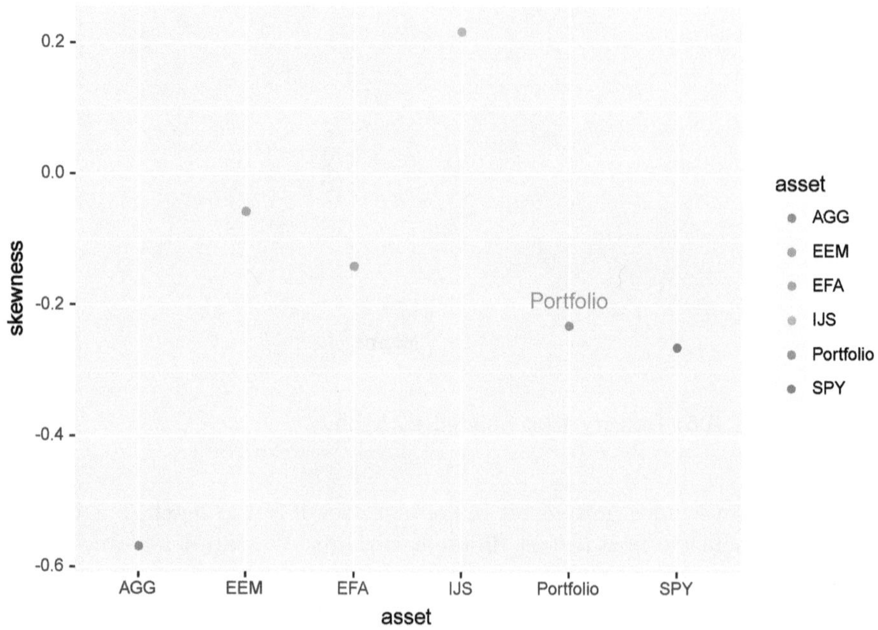

FIGURE 5.6: Asset and Portfolio Skewness Comparison

Figure 5.6 indicates that our portfolio has a lower skewness than all of the individual assets except SPY and AGG.

5.4   Rolling Skewness in the xts world

For the same reasons that we did so with standard deviation, let's check whether we have missed anything unusual in the portfolio's historical tail risk by examining rolling skewness.

In the xts world, calculating rolling skewness is almost identical to calculating rolling standard deviation, except we call the skewness() function instead of

`StdDev()`. Since this is a rolling calculation, we need a period of time and will use a 24-month window.

```
window <- 24
rolling_skew_xts <-
  rollapply(portfolio_returns_xts_rebalanced_monthly,
            FUN = skewness,
            width = window) %>%
  na.omit()
```

5.5 Rolling Skewness in the tidyverse with tibbletime

As we saw with standard deviation, passing a rolling calculation to `dplyr` pipes does not work smoothly. We can, though, use `rollify()` from `tibbletime`.

We first create a rolling function.

```
skew_roll_24 <-
  rollify(skewness, window = window)
```

We then convert our portfolio returns to a `tibbletime` object and pass them to the rolling function.

```
roll_skew_tibbletime <-
  portfolio_returns_tq_rebalanced_monthly %>%
  as_tbl_time(index = date) %>%
  mutate(skew = skew_roll_24(returns)) %>%
  select(-returns) %>%
  na.omit()
```

5.6 Rolling Skewness in the tidyquant world

In the `tidyquant` world, we wrap `rollapply()` within `tq_mutate`, and then supply the `skewness()` function.

```
rolling_skew_tq <-
  portfolio_returns_tq_rebalanced_monthly %>%
  tq_mutate(select = returns,
            mutate_fun = rollapply,
            width      = window,
            FUN        = skewness,
            col_rename = "tq") %>%
  na.omit()
```

Let's confirm that our results are consistent and then start visualizing.

```
rolling_skew_tq %>%
  select(-returns) %>%
  mutate(xts = coredata(rolling_skew_xts),
         tbltime = roll_skew_tibbletime$skew) %>%
  mutate_if(is.numeric, funs(round(., 3))) %>%
  tail(3)
```

```
# A tibble: 3 x 4
  date                tq       xts   tbltime
  <date>           <dbl>     <dbl>     <dbl>
1 2017-10-31    0.0830    0.0830    0.0830
2 2017-11-30   -0.00400  -0.00400  -0.00400
3 2017-12-31    0.0180    0.0180    0.0180
```

5.7 Visualizing Rolling Skewness

Our visualization flow for skewness is quite similar to our work on standard
deviation. We start by passing `rolling_skew_xts` into `highcharter`. We also
tweak our y-axis to capture the nature of the rolling fluctuations by setting
the range to between 2 and -2 with `hc_yAxis(..., max = 2, min = -2)`.

```
highchart(type = "stock") %>%
  hc_title(text = "Rolling 24-Month Skewness") %>%
  hc_add_series(rolling_skew_xts,
                name = "Rolling skewness",
                color = "cornflowerblue") %>%
  hc_yAxis(title = list(text = "skewness"),
           opposite = FALSE,
```

```
                max = 1,
                min = -1)   %>%
    hc_navigator(enabled = FALSE) %>%
    hc_scrollbar(enabled = FALSE) %>%
    hc_add_theme(hc_theme_flat()) %>%
    hc_exporting(enabled = TRUE)
```

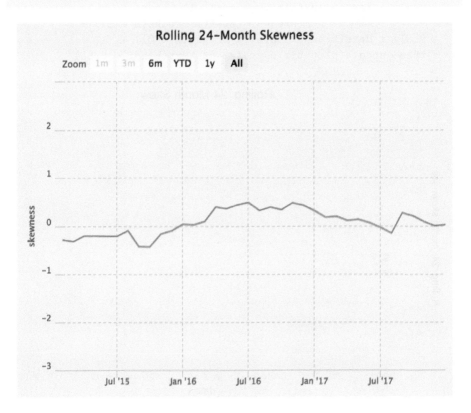

FIGURE 5.7: Rolling Skewness highcharter

Figure 5.7 shows the movement in rolling skewness, try re-running the code without enforcing limits on the y-axis.

We create a similar visualization with `ggplot()` and our `rolling_skew_tq` object.

I will again impose minimum and maximum y-axis values, with `scale_y_continuous(limits = c(-1, 1)...)`.

```
rolling_skew_tq %>%
  ggplot(aes(x = date, y = tq)) +
  geom_line(color = "cornflowerblue") +
  ggtitle("Rolling  24-Month Skew ") +
  ylab(paste("Rolling ", window, " month skewness",
             sep = " ")) +
  scale_y_continuous(limits = c(-1, 1),
                        breaks = pretty_breaks(n = 8)) +
  scale_x_date(breaks = pretty_breaks(n = 8)) +
  theme_update(plot.title = element_text(hjust = 0.5))
```

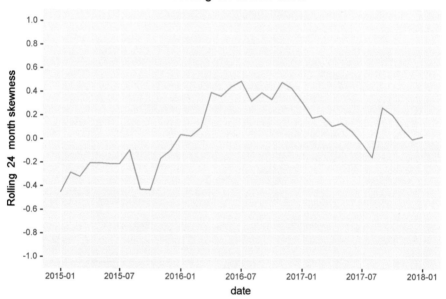

FIGURE 5.8: Rolling Skewness ggplot

Figure 5.8 makes the rolling skewness seem more volatile than Figure 5.7. Tweaking the y-axis can have a big effect, use it wisely.

The rolling charts are quite illuminating and show that the 24-month skewness has been positive for about half the lifetime of this portfolio even though the overall skewness is negative. Normally we would now head to Shiny and enable a way to test different rolling windows but let's wait until we cover kurtosis in the next chapter.

6

Kurtosis

Kurtosis is a measure of the degree to which portfolio returns appear in the tails of their distribution. A normal distribution has a kurtosis of 3, which follows from the fact that a normal distribution does have some of its mass in its tails. A distribution with a kurtosis greater than 3 has more returns in its tails than the normal, and one with kurtosis less than 3 has fewer returns in its tails than the normal. That matters to investors because more bad returns in the tails means that our portfolio might be at risk of a rare but huge downside event. The terminology is a bit confusing because negative kurtosis actually is less risky because it has fewer returns in the tails.

Kurtosis is often described as negative excess or positive excess, and that is in comparison to a kurtosis of 3. A distribution with negative excess kurtosis equal to -1 has an absolute kurtosis of 2, but we subtract 3 from 2 to get to -1. Remember, though, the negative kurtosis means fewer returns in the tails and, probably, less risk.

Here is the equation for excess kurtosis. Note that we subtract 3 at the end:

$$Kurtosis = \sum_{t=1}^{n}(x_i - \overline{x})^4/n \left/ \left(\sum_{t=1}^{n}(x_i - \overline{x})^2/n\right)^2 - 3\right.$$

The code flows for calculating kurtosis and rolling kurtosis are quite similar to those for skewness, except we use the built-in **kurtosis()** function. That was by design, we want to write code that can easily be reused for another project.

6.1   Kurtosis in the xts world

For the xts world, we use the **kurtosis()** function instead of the **skewness()** function.

```
kurt_xts <-
  kurtosis(portfolio_returns_xts_rebalanced_monthly$returns)
```

6.2   Kurtosis in the tidyverse

For tidy, we have the same piped flow and use the formula for kurtosis for our
by-hand calculations.

```
kurt_tidy <-
  portfolio_returns_tq_rebalanced_monthly %>%
  summarise(
  kurt_builtin = kurtosis(returns),
  kurt_byhand =
  ((sum((returns - mean(returns))^4)/
      length(returns))/
  ((sum((returns - mean(returns))^2)/
      length(returns))^2)) - 3) %>%
  select(kurt_builtin, kurt_byhand)
```

We confirm that we have consistent calculations.

```
kurt_tidy %>%
  mutate(xts = kurt_xts)
```

```
# A tibble: 1 x 3
  kurt_builtin kurt_byhand    xts
         <dbl>       <dbl>  <dbl>
1        0.457       0.457  0.457
```

We were able to reuse our code from above to shorten the development time
here. We will do the same with visualizations.

6.3   Visualizing Kurtosis

We already have the `portfolio_density_plot` object and can customize the
visualizations for kurtosis instead of skewness. We are now more interested in

both tails, so we shade at 2 standard deviations above and below the mean return (for our skewness work, we shaded the negative tail).

```
sd_pos <-
  mean + (2* sd(portfolio_returns_tq_rebalanced_monthly$returns))
sd_neg <-
  mean - (2* sd(portfolio_returns_tq_rebalanced_monthly$returns))

sd_pos_shaded_area <-
  ggplot_build(portfolio_density_plot)$data[[1]] %>%
  filter(x > sd_pos )

sd_neg_shaded_area <-
  ggplot_build(portfolio_density_plot)$data[[1]] %>%
  filter(x < sd_neg)

  portfolio_density_plot +
  geom_area(data = sd_pos_shaded_area,
            aes(x = x, y = y),
            fill="pink",
            alpha = 0.5) +
  geom_area(data = sd_neg_shaded_area,
            aes(x = x, y = y),
            fill="pink",
            alpha = 0.5) +
  scale_x_continuous(breaks = pretty_breaks(n = 10))
```

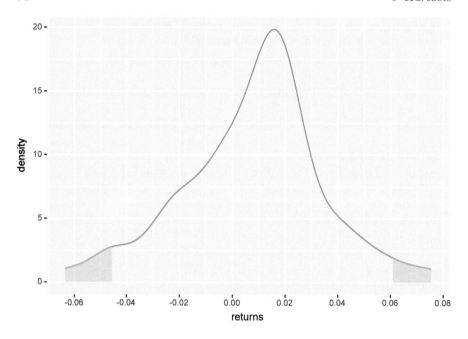

FIGURE 6.1: Kurtosis Density ggplot

Figure 6.1 looks good, but we can add more aesthetics.

```
portfolio_density_plot +
  geom_area(data = sd_pos_shaded_area,
            aes(x = x, y = y),
            fill="pink",
            alpha = 0.5) +
  geom_area(data = sd_neg_shaded_area,
            aes(x = x, y = y),
            fill="pink",
            alpha = 0.5) +
  geom_segment(data = shaded_area_data,
               aes(x = mean,
                   y = 0,
                   xend = mean,
                   yend = density),
               color = "red",
               linetype = "dotted") +

  annotate(geom = "text",
           x = mean,
```

```
            y = 5,
            label = "mean",
            color = "red",
            fontface = "plain",
            angle = 90,
            alpha = .8,
            vjust =  -1.75) +

geom_segment(data = median_line_data,
            aes(x = median,
                y = 0,
                xend = median,
                yend = density),
            color = "black",
            linetype = "dotted") +

annotate(geom = "text",
        x = median,
        y = 5,
        label = "median",
        fontface = "plain",
        angle = 90,
        alpha = .8,
        vjust =  1.75) +
scale_x_continuous(breaks = pretty_breaks(n = 10))
```

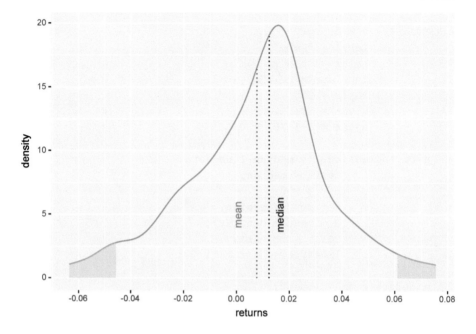

FIGURE 6.2: Kurtosis Density ggplot with Lines

Figure 6.2 is a good example of a ggplot() chart that we might want to save somewhere as a template. It contains several aesthetics that could come in handy for future work.

Let's also compare our portfolio kurtosis to the individual assets' kurtosis.

```
asset_returns_long %>%
  summarize(kurt_assets = kurtosis(returns)) %>%
  add_row(asset = "Portfolio",
          kurt_assets = kurt_tidy$kurt_byhand) %>%
  ggplot(aes(x = asset,
             y = kurt_assets,
             colour = asset)) +
  geom_point() +
  geom_text(
        aes(x = "Portfolio",
            y =
               kurt_tidy$kurt_byhand + .06),
            label = "Portfolio",
          color = "cornflowerblue") +
  labs(y = "kurtosis")
```

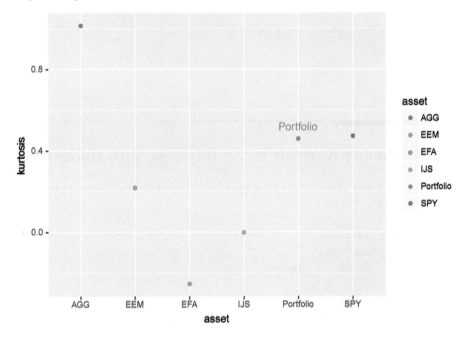

FIGURE 6.3: Asset and Portfolio Kurtosis Comparison

Figure 6.3 should look familiar, it's the same format as our colored chart for standard deviation and skewness.

Building those kurtosis visualizations was much smoother because we could rely on our previous work, and the same will be true for rolling kurtosis.

6.4 Rolling Kurtosis in the xts world

Calculating rolling kurtosis in the `xts` world uses the same code flow as we used for skewness, except we replace `FUN = skewness` with `FUN = kurtosis`.

```
window <- 24

rolling_kurt_xts <-
  rollapply(portfolio_returns_xts_rebalanced_monthly,
            FUN = kurtosis,
```

```
                  width = window) %>%
    na.omit()
```

6.5   Rolling Kurtosis in the tidyverse with tibbletime

In the tidyverse plus tibbletime paradigm, we return to the rollify() code
flow.

```
kurt_roll_24 <-
    rollify(kurtosis,
            window = window)

roll_kurt_tibbletime <-
    portfolio_returns_tq_rebalanced_monthly %>%
    as_tbl_time(index = date) %>%
    mutate(kurt = kurt_roll_24(returns)) %>%
    select(-returns) %>%
    na.omit()
```

6.6   Rolling Kurtosis in the tidyquant world

In the tidyquant world, we, again, wrap rollapply with tq_mutate()and
then call FUN = kurtosis.

```
rolling_kurt_tq <-
    portfolio_returns_tq_rebalanced_monthly %>%
    tq_mutate(select = returns,
              mutate_fun = rollapply,
              width      = window,
              FUN        = kurtosis,
              col_rename = "tq") %>%
    select(-returns) %>%
    na.omit()
```

Let's compare our results.

```
rolling_kurt_tq %>%
  mutate(xts = coredata(rolling_kurt_xts),
         tbltime = roll_kurt_tibbletime$kurt) %>%
  mutate_if(is.numeric, funs(round(.,3))) %>%
  tail(3)
```

```
# A tibble: 3 x 4
  date             tq    xts  tbltime
  <date>        <dbl>  <dbl>    <dbl>
1 2017-10-31     2.13   2.13     2.13
2 2017-11-30     2.22   2.22     2.22
3 2017-12-31     3.38   3.38     3.38
```

6.7 Visualizing Rolling Kurtosis

We can pop our `rolling_kurt_xts` object into `highcharter` for visualization, same as we did for skewness.

```
highchart(type = "stock") %>%
  hc_title(text = "Rolling 24-Month kurtosis") %>%
  hc_add_series(rolling_kurt_xts,
                name = "Rolling 24-Month kurtosis",
                color = "cornflowerblue") %>%
  hc_yAxis(title = list(text = "kurtosis"),
           opposite = FALSE) %>%
  hc_add_theme(hc_theme_flat()) %>%
  hc_navigator(enabled = FALSE) %>%
  hc_scrollbar(enabled = FALSE) %>%
  hc_exporting(enabled = TRUE)
```

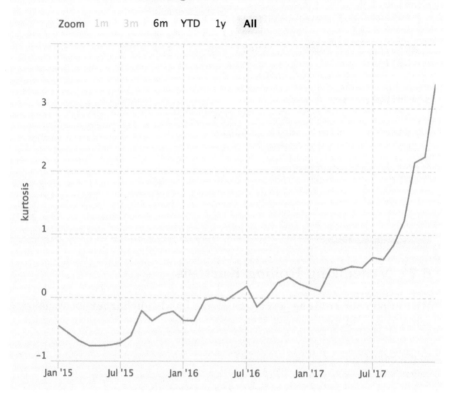

FIGURE 6.4: Rolling Kurtosis highcharter

Figure 6.4 shows an increasing rolling kurtosis since 2015.

Let's complete our rolling kurtosis visualization with `ggplot()`.

```
rolling_kurt_tq %>%
  ggplot(aes(x = date, y = tq)) +
  geom_line(color = "cornflowerblue") +
  scale_y_continuous(breaks = pretty_breaks(n = 8)) +
  scale_x_date(breaks = pretty_breaks(n = 8)) +
  ggtitle("Rolling 24-Month Kurtosis") +
  labs(y = "rolling kurtosis") +
  theme_update(plot.title = element_text(hjust = 0.5))
```

FIGURE 6.5: Rolling Kurtosis ggplot

Figure 6.5 looks how we were expecting since it's the same data as we used for Figure 6.4.

That's all for our work on kurtosis, which was made a lot more efficient by our work on skewness. Now let's create one app for interactively visualizing both skewness and kurtosis.

6.8 Shiny App Skewness and Kurtosis

To wrap our skewness and kurtosis work into a Shiny application, we start with the same sidebar for stocks, weights, starting date and rolling window, as shown in Figure 6.6.

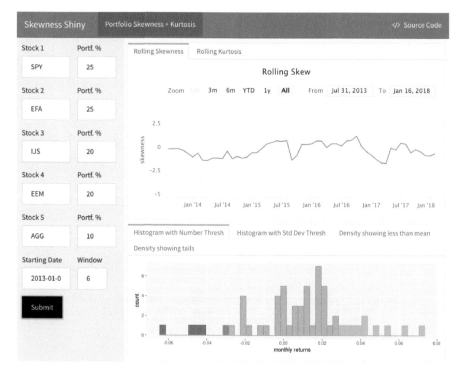

FIGURE 6.6: www.reproduciblefinance.com/shiny/skewness-kurtosis

Let's get into the code for this app.

There are three crucial `eventReactive()` chunks.

First, we calculate portfolio returns based on user input and stay in the `xts` world.

```
portfolio_returns_xts <- eventReactive(input$go, {
  prices <- prices()

  w <- c(input$w1/100, input$w2/100,
         input$w3/100, input$w4/100,
         input$w5/100)

  prices_monthly <-
    to.monthly(prices, indexAt = "last", OHLC = FALSE)

  asset_returns_xts <-
    na.omit(Return.calculate(prices_monthly, method = "log"))
```

```
  portfolio_returns_xts <-
    Return.portfolio(asset_returns_xts, weights = w) %>%
    `colnames<-`("returns")
})
```

Then we pass `portfolio_returns_xts()` to our first `eventReactive()` to calculate rolling skewness.

```
rolling_skew_xts <- eventReactive(input$go, {
  rolling_skew_xts <-
    rollapply(portfolio_returns_xts(),
          FUN = skewness,
          width = input$window) %>%
    na.omit()
})
```

Next we pass `portfolio_returns_xts()` to our second `eventReactive()` to calculate rolling kurtosis.

```
rolling_kurt_xts <- eventReactive(input$go, {
  rolling_kurt_xts <-
    rollapply(portfolio_returns_xts(),
          FUN = kurtosis,
          width = input$window) %>%
    na.omit()
})
```

We now have the objects `rolling_skew_xts` and `rolling_kurt_xts` and can pass them directly to `highcharter`.

We display rolling skewness with `hc_add_series(rolling_skew_xts(),....` Note that we start with `renderHighchart()` to alert Shiny that a reactive `highcharter` is about to be built.

```
renderHighchart({

  highchart(type = "stock") %>%
    hc_title(text = "Rolling Skew") %>%
    hc_add_series(rolling_skew_xts(),
                  name = "rolling skew",
                  color = "cornflowerblue") %>%
    hc_yAxis(title = list(text = "skewness"),
           opposite = FALSE,
```

```
            max = 3,
            min = -3) %>%
  hc_navigator(enabled = FALSE) %>%
  hc_scrollbar(enabled = FALSE)  %>%
  hc_exporting(enabled = TRUE)

})
```

We display rolling kurtosis with `hc_add_series(rolling_kurt_xts(),....`

```
renderHighchart({

  highchart(type = "stock") %>%
    hc_title(text = "Rolling Kurtosis") %>%
    hc_add_series(rolling_kurt_xts(),
                    name = "rolling kurt",
                    color = "cornflowerblue") %>%
    hc_yAxis(title = list(text = "kurtosis"),
             opposite = FALSE,
             max = 3,
             min = -3) %>%
  hc_navigator(enabled = FALSE) %>%
  hc_scrollbar(enabled = FALSE) %>%
  hc_exporting(enabled = TRUE)

})
```

The app also includes two `ggplot()` visualizations. The code flows for these are the same as we reviewed in the sections on visualizing skewness and kurtosis, except they operate on a reactive data object, `portfolio_returns_tq_rebalanced_monthly()`.

First, there is a reactive histogram of returns, shaded by distance from the mean.

```
renderPlot({

  portfolio_returns_tq_rebalanced_monthly() %>%
  mutate(hist_col_red =
             ifelse(returns < (mean(returns) - 2*sd(returns)),
                    returns, NA),
           hist_col_green =
             ifelse(returns > (mean(returns) + 2*sd(returns)),
                    returns, NA),
```

```
          hist_col_blue =
            ifelse(returns > (mean(returns) - 2*sd(returns)) &
                   returns < (mean(returns) + 2*sd(returns)),
                   returns, NA)) %>%
  ggplot() +

  geom_histogram(aes(x = hist_col_red),
              alpha = .7,
              binwidth = .003,
              fill = "red",
              color = "red") +

  geom_histogram(aes(x = hist_col_green),
              alpha = .7,
              binwidth = .003,
              fill = "green",
              color = "green") +

  geom_histogram(aes(x = hist_col_blue),
              alpha = .7,
              binwidth = .003,
              fill = "cornflowerblue",
              color = "cornflowerblue") +

scale_x_continuous(breaks = pretty_breaks(n = 10)) +
xlab("monthly returns")
})
```

Next, there is a reactive density chart, shaded by returns below the mean.

```
renderPlot({

  portfolio_returns_tq_rebalanced_monthly <-
    portfolio_returns_tq_rebalanced_monthly()

  mean <-
    mean(portfolio_returns_tq_rebalanced_monthly$returns)

  median <-
    median(portfolio_returns_tq_rebalanced_monthly$returns)

  skew_density_plot <-
    portfolio_returns_tq_rebalanced_monthly %>%
```

```r
    ggplot(aes(x = returns)) +
    stat_density(geom = "line",
                 size = 1,
                 color = "cornflowerblue")

  shaded_area_data <-
    ggplot_build(skew_density_plot)$data[[1]] %>%
    filter(x < mean)

  skew_density_plot_shaded <-
    skew_density_plot +
    geom_area(data = shaded_area_data,
              aes(x = x, y = y),
              fill="pink",
              alpha = 0.5)

  median_line_data <-
    ggplot_build(skew_density_plot)$data[[1]] %>%
    filter(x <= median)

skew_density_plot_shaded +

  geom_segment(data = median_line_data,
               aes(x = median, y = 0,
                   xend = median,
                   yend = density),
               color = "black",
               linetype = "dotted") +

  annotate(geom = "text",
           x = median,
           y = 5,
           label = "median",
           fontface = "plain",
           angle = 90,
           alpha = .8,
           vjust = 1.75) +

  annotate(geom = "text",
           x = (mean - .03),
           y = .1,
           label = "returns < mean",
           fontface = "plain",
           color = "red",
```

```
            alpha = .8,
            vjust =  -1) +
   ggtitle("Density Plot Illustrating Skewness")
})
```

Our Shiny app for skewness and kurtosis is now complete. Along with the standard deviation app, an end user now has ample options for building a portfolio and exploring the dispersion of returns around the mean. Consider how we could combine elements of all these apps to form one returns dispersion inspection tool.

Concluding Risk

That completes our section on risk, which we treated as the variability of portfolio returns. We have explored several code flows for measuring historical and rolling standard deviation, skewness and kurtosis. From a data science toolkit perspective, we have delved into descriptive statistics for our portfolio, focusing on the variability or dispersion of returns.

One important take-away from this section is how we reused our own code flows to accelerate future work. Our work on skewness was facilitated by our work on standard deviation, and our work on kurtosis flowed from the work on skewness. Writing clear, reproducible code in the first chapter might have taken us a bit more time up front, but it had a nice efficiency payoff in the future chapters. When our work becomes more complex in the real world, those future efficiencies become ever more important. Hopefully this chapter has convinced us that writing good code is not simply an aesthetic nice-to-have - it has tangible benefits by saving us time.

If you are starting a new R session and wish to run our code for the different risk measures calculated in this section, first get the data objects:

www.reproduciblefinance.com/code/get-returns/

And then see these pages:

www.reproduciblefinance.com/code/standard-deviation/

www.reproduciblefinance.com/code/skewness/

www.reproduciblefinance.com/code/kurtosis/

Portfolio Theory

In this section we devote three chapters to the relationship between risk and return. These topics are the most theoretical that we have covered yet, but we will not be delving into the theory. Instead, we will focus on code flows.

First, we will discuss the Sharpe Ratio, a measure of the return versus risk ratio of a portfolio.

Then, we will look at the Capital Asset Pricing Model (CAPM) and specifically how to calculate the market beta for our assets and portfolio. This will be an introduction to simple linear regression.

We will conclude with an exploration of the Fama-French multi-factor model, which also serves as an introduction to multiple linear regression.

If you wish to study further into these topics, see Sharpe's 1964 article, "Asset Prices: A Theory of Market Equilibrium under Conditions of Risk",[1] Sharpe's 1994 article "The Sharpe Ratio",[2] and "Common risk factors in the returns on stocks and bonds"[3] by Fama and French.

From a general data science perspective, we have covered data import and wrangling in the first section, descriptive statistics in the second section, and this section is devoted to the modeling and evaluating of our data.

We will accomplish the following in this section:

1) calculate and visualize the Sharpe Ratio and the rolling Sharpe Ratio
2) calculate and visualize CAPM beta
3) calculate and visualize the Fama-French 3-Factor Model and the rolling Fama-French 3-Factor model
4) build Shiny apps for Sharpe Ratio, CAPM beta and rolling Fama-French model

We will be working with the portfolio returns objects that were created in the

---

[1]Sharpe, William F.. (1964). "Asset Prices: A Theory of Market Equilibrium under Conditions of Risk". The Journal of Finance, Vol. 19, No. 3 pp. 425-442.

[2]Sharpe, William F. (1994). "The Sharpe Ratio". The Journal of Portfolio Management. 21 (1): 49–58.

[3]Fama, Eugene and French, Kenneth. "Common risk factors in the returns on stocks and bonds" Journal of Financial Economics Volume 33, Issue 1, February 1993, Pages 3-56.

Returns section. If you are starting a new R session and want to run the code to build those objects, navigate here:

www.reproduciblefinance.com/code/get-returns/

7

Sharpe Ratio

The Sharpe Ratio is defined as the mean of the excess monthly portfolio returns above the risk-free rate, divided by the standard deviation of the excess monthly portfolio returns above the risk-free rate. This is the formulation of the Sharpe Ratio as of 1994; if we wished to use the original formulation from 1966 the denominator would be the standard deviation of all the monthly portfolio returns.

The Sharpe Ratio measures excess returns per unit of risk, where we again take the standard deviation to represent portfolio risk. The Sharpe Ratio was brought to us by Bill Sharpe - arguably the most important economist for modern investment management as the creator of the Sharpe Ratio, CAPM (which we will cover later) and Financial Engines, a forerunner of today's robo-advisor movement.

The Sharpe Ratio equation is as follows:

$$Sharpe\ Ratio = (\overline{R_p - R_f})/\sigma_{excess}$$

The numerator is the mean excess return above the risk-free rate and the denominator is the standard deviation of those excess returns. In other words, it is the ratio of return to risk and so a higher Sharpe Ratio indicates a 'better' portfolio.

We will start with the built-in function from the xts world and will look at the by-hand equation as part of the tidyverse.

7.1 Sharpe Ratio in the xts world

For any work with the Sharpe Ratio, we first must choose a risk-free rate (hereafter RFR) and will use .3%.

```
rfr <- .0003
```

From there, calculating the Sharpe Ratio in the xts world is almost depressingly convenient. We call SharpeRatio(portfolio_returns_xts, Rf = rfr), passing our portfolio returns and risk-free rate to the built-in function from PerformanceAnalytics.

```
sharpe_xts <-
  SharpeRatio(portfolio_returns_xts_rebalanced_monthly,
             Rf = rfr,
             FUN = "StdDev") %>%
  `colnames<-`("sharpe_xts")

sharpe_xts
```

```
                             sharpe_xts
StdDev Sharpe (Rf=0%, p=95%):    0.2749
```

7.2   Sharpe Ratio in the tidyverse

For our tidyverse example, we will implement the Sharpe Ratio equation via pipes and dplyr.

We start with our object portfolio_returns_dplyr_byhand and then run summarise(ratio = mean(returns - rfr)/sd(returns - rfr)),   which maps to the equation for the Sharpe Ratio.

```
sharpe_tidyverse_byhand <-
  portfolio_returns_dplyr_byhand %>%
  summarise(sharpe_dplyr = mean(returns - rfr)/
           sd(returns - rfr))

sharpe_tidyverse_byhand
```

```
# A tibble: 1 x 1
  sharpe_dplyr
         <dbl>
1        0.275
```

7.3   Sharpe Ratio in the tidyquant world

tidyquant allows us to wrap the SharpeRatio() function inside the tq_performance() function.

```
sharpe_tq <-
  portfolio_returns_tq_rebalanced_monthly %>%
  tq_performance(Ra = returns,
                 performance_fun = SharpeRatio,
                 Rf = rfr,
                 FUN = "StdDev") %>%
  `colnames<-`("sharpe_tq")
```

Let's compare our 3 Sharpe objects.

```
sharpe_tq %>%
  mutate(tidy_sharpe = sharpe_tidyverse_byhand$sharpe_dplyr,
         xts_sharpe = sharpe_xts)
```

```
# A tibble: 1 x 3
  sharpe_tq tidy_sharpe xts_sharpe
      <dbl>       <dbl>      <dbl>
1     0.275       0.275      0.275
```

We have consistent results from xts, tidyquant and our by-hand piped calculation. Next, we compare to the Sharpe Ratio of the S&P500 in the same time period.

```
market_returns_xts <-
    getSymbols("SPY",
               src = 'yahoo',
               from = "2012-12-31",
               to = "2017-12-31",
               auto.assign = TRUE,
               warnings = FALSE) %>%
    map(~Ad(get(.))) %>%
    reduce(merge) %>%
    `colnames<-`("SPY") %>%
    to.monthly(indexAt = "lastof",
               OHLC = FALSE)

market_sharpe <-
```

```
market_returns_xts %>%
tk_tbl(preserve_index = TRUE,
        rename_index = "date") %>%
  mutate(returns =
              (log(SPY) - log(lag(SPY)))) %>%
  na.omit() %>%
  summarise(ratio =
              mean(returns - rfr)/sd(returns - rfr))
```

```
market_sharpe$ratio
```

`[1] 0.4348`

Our portfolio has *underperformed* the market during our chosen time period. Welcome to the challenges of portfolio construction during a raging bull market.

7.4   Visualizing Sharpe Ratio

Before visualizing the actual Sharpe, we will get a sense for what proportion of our portfolio returns exceeded the RFR.

When we originally calculated Sharpe by-hand in the tidyverse, we used `summarise` to create one new cell for our end result. The code was `summarise(ratio = mean(returns - rfr)/sd(returns - rfr))`.

Now, we will make two additions to assist in our data visualization. We will add a column for returns that fall below the risk-free rate with `mutate(returns_below_rfr = ifelse(returns < rfr, returns, NA))` and add a column for returns above the risk-free rate with `mutate(returns_above_rfr = ifelse(returns > rfr, returns, NA))`.

This is not necessary for calculating the Sharpe Ratio, but we will see how it illustrates a benefit of doing things by-hand with `dplyr`: if we want to extract or create certain data transformations, we can add it to the piped code flow.

```
sharpe_byhand_with_return_columns <-
  portfolio_returns_tq_rebalanced_monthly %>%
  mutate(ratio =
          mean(returns - rfr)/sd(returns - rfr)) %>%
  mutate(returns_below_rfr =
          if_else(returns < rfr, returns, as.numeric(NA))) %>%
```

```
  mutate(returns_above_rfr =
          if_else(returns > rfr, returns, as.numeric(NA))) %>%
  mutate_if(is.numeric, funs(round(.,4)))

sharpe_byhand_with_return_columns %>%
  head(3)
```

```
# A tibble: 3 x 5
  date           returns ratio returns_below_rfr
  <date>           <dbl> <dbl>             <dbl>
1 2013-01-31    0.0308   0.275               NA
2 2013-02-28   -0.000900 0.275         -0.000900
3 2013-03-31    0.0187   0.275               NA
# ... with 1 more variable: returns_above_rfr <dbl>
```

Now we can create a scatter plot in order to quickly grasp how many of our returns are above the RFR and how many are below the RFR.

We will create green points for returns above RFR with geom_point(aes(y = returns_above_RFR), colour = "green") and red points for returns below RFR with geom_point(aes(y = returns_below_rfr), colour = "red").

We also add a blue vertical line at November of 2016 for the election and a horizontal purple dotted line at the RFR.

```
sharpe_byhand_with_return_columns %>%
  ggplot(aes(x = date)) +
  geom_point(aes(y = returns_below_rfr),
             colour = "red") +
  geom_point(aes(y = returns_above_rfr),
             colour = "green") +
  geom_vline(xintercept =
                as.numeric(as.Date("2016-11-30")),
             color = "blue") +
  geom_hline(yintercept = rfr,
             color = "purple",
             linetype = "dotted") +
  annotate(geom = "text",
           x = as.Date("2016-11-30"),
           y = -.04,
           label = "Election",
           fontface = "plain",
           angle = 90,
           alpha = .5,
           vjust = 1.5) +
```

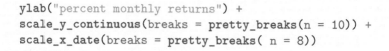

```
  ylab("percent monthly returns") +
  scale_y_continuous(breaks = pretty_breaks(n = 10)) +
  scale_x_date(breaks = pretty_breaks( n = 8))
```

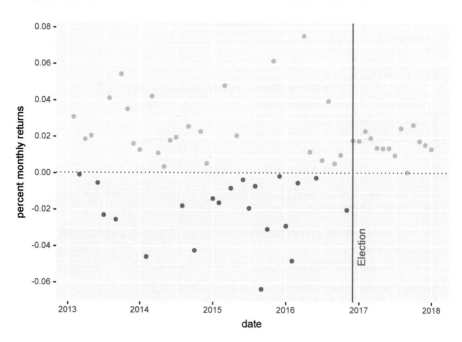

FIGURE 7.1: Scatter Returns Around Risk Free Rate

Have a look at Figure 7.1 and notice that there are zero returns below the RFR after the election in 2016.

Next we will build a histogram of the distribution of returns with `geom_histogram(alpha = 0.25, binwidth = .01, fill = "cornflowerblue")` and add a vertical line at the RFR.

```
sharpe_byhand_with_return_columns %>%
  ggplot(aes(x = returns)) +
  geom_histogram(alpha = 0.45,
                 binwidth = .01,
                 fill = "cornflowerblue") +
  geom_vline(xintercept = rfr,
             color = "green") +
  annotate(geom = "text",
           x = rfr,
```

```
          y = 13,
          label = "rfr",
          fontface = "plain",
          angle = 90,
          alpha = .5,
          vjust = 1)
```

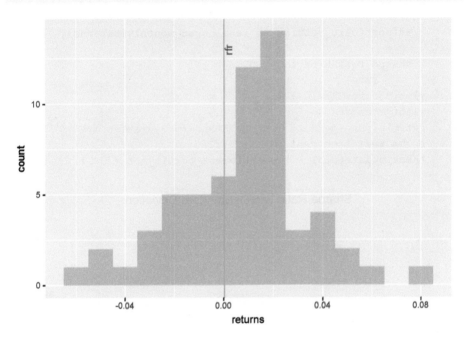

FIGURE 7.2: Returns Histogram with Risk-Free Rate ggplot

Figure 7.2 shows the distribution of returns in comparison to the risk-free rate, but we have not visualized the actual Sharpe Ratio yet.

Similar to standard deviation, skewness and kurtosis of returns, it helps to visualize the Sharpe Ratio of our portfolio in comparison to other assets.

```
asset_returns_long %>%
  summarise(stand_dev = sd(returns),
            sharpe = mean(returns - rfr)/
              sd(returns - rfr))%>%
  add_row(asset = "Portfolio",
    stand_dev =
      portfolio_sd_xts_builtin[1],
```

```
    sharpe =
       sharpe_tq$sharpe_tq) %>%
 ggplot(aes(x = stand_dev,
             y = sharpe,
             color = asset)) +
 geom_point(size = 2) +
 geom_text(
  aes(x =
   sd(portfolio_returns_tq_rebalanced_monthly$returns),
    y =
   sharpe_tq$sharpe_tq + .02,
       label = "Portfolio")) +
 ylab("Sharpe Ratio") +
 xlab("standard deviation") +
 ggtitle("Sharpe Ratio versus Standard Deviation") +
 # The next line centers the title
 theme_update(plot.title = element_text(hjust = 0.5))
```

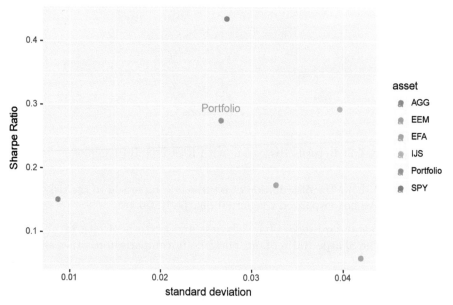

FIGURE 7.3: Sharpe versus Standard Deviation

Figure 7.3 indicates that the S&P500 again seems to dominate our portfolio, though it does have slightly more risk.

That's interesting to observe but, just as with standard deviation, skewness and kurtosis, these overall numbers might obscure important periods of fluctuation in our data. We can solve that by working with the rolling Sharpe Ratio.

7.5 Rolling Sharpe Ratio in the xts world

Very similar to how we calculated rolling standard deviation, skewness and kurtosis, our xts work starts with `rollapply()`.

Note that we use a more wordy function format here because we need to pass in the argument `FUN = "StdDev"`. Try running the code without that argument and review the error.

```
window <- 24

rolling_sharpe_xts <-
  rollapply(portfolio_returns_xts_rebalanced_monthly,
            window,
            function(x)
            SharpeRatio(x,
                        Rf = rfr,
                        FUN = "StdDev")) %>%
  na.omit() %>%
  `colnames<-`("xts")
```

7.6 Rolling Sharpe Ratio with the tidyverse and tibbletime

We can combine the tidyverse and `tibbletime` to calculate the rolling Sharpe Ratio calculation starting from a `tibble`.

We first write our own function by combining `rollify()` and `ratio = mean(returns - rfr)/sd(returns - rfr)`.

Notice we still pass in our `rfr` and `window` variables from previous code chunks.

```
# Creat rolling function.
sharpe_roll_24 <-
  rollify(function(returns) {
    ratio = mean(returns - rfr)/sd(returns - rfr)
    },
window = window)
```

Next we pass our portfolio data object to that rolling function, via `mutate()`.

```
rolling_sharpe_tidy_tibbletime <-
  portfolio_returns_dplyr_byhand %>%
  as_tbl_time(index = date) %>%
  mutate(tbltime_sharpe = sharpe_roll_24(returns)) %>%
  na.omit() %>%
  select(-returns)
```

7.7   Rolling Sharpe Ratio with tidyquant

To calculate the rolling Sharpe Ratio with `tidyquant`, we first build a custom function where we can specify the RFR and an argument to the `SharpeRatio()` function. Again, our rolling Sharpe Ratio work is more complex than previous rolling calculations.

```
sharpe_tq_roll <- function(df){
  SharpeRatio(df,
              Rf = rfr,
              FUN = "StdDev")
}
```

It is necessary to build that custom function because we would not be able to specify `FUN = "StdDev"` otherwise.

Now we use `tq_mutate()` to wrap `rollapply()` and our custom function, and apply them to `portfolio_returns_tq_rebalanced_monthly`.

```
rolling_sharpe_tq <-
portfolio_returns_tq_rebalanced_monthly %>%
tq_mutate(
        select      = returns,
```

```
          mutate_fun = rollapply,
          width      = window,
          align      = "right",
          FUN        = sharpe_tq_roll,
          col_rename = "tq_sharpe"
    ) %>%
  na.omit()
```

Now we can compare our 3 rolling Sharpe Ratio objects and confirm consistency.

```
rolling_sharpe_tidy_tibbletime %>%
  mutate(xts_sharpe = coredata(rolling_sharpe_xts),
         tq_sharpe = rolling_sharpe_tq$tq_sharpe ) %>%
  head(3)
```

```
# A time tibble: 3 x 4
# Index: date
  date         tbltime_sharpe xts_sharpe tq_sharpe
  <date>              <dbl>      <dbl>      <dbl>
1 2014-12-31          0.312      0.312      0.312
2 2015-01-31          0.237      0.237      0.237
3 2015-02-28          0.300      0.300      0.300
```

7.8 Visualizing the Rolling Sharpe Ratio

Finally, we can start to visualize the Sharpe Ratio across the history of the portfolio.

We start with `highcharter` and `xts`.

```
highchart(type = "stock") %>%
  hc_title(text = "Rolling  24-Month Sharpe") %>%
  hc_add_series(rolling_sharpe_xts,
                name = "sharpe",
                color = "blue") %>%
  hc_navigator(enabled = FALSE) %>%
  hc_scrollbar(enabled = FALSE) %>%
  hc_add_theme(hc_theme_flat()) %>%
  hc_exporting(enabled = TRUE)
```

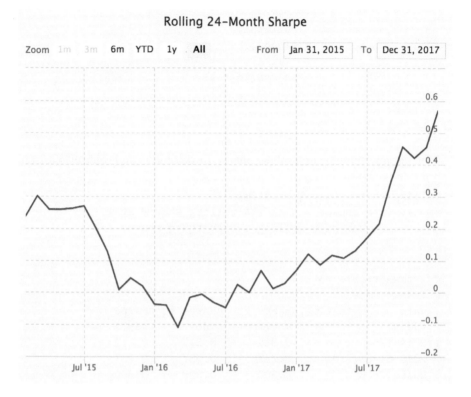

FIGURE 7.4: Rolling Sharpe highcharter

Figure 7.4 is confirming a trend that we noticed previously, that this portfolio has done quite well since November of 2016.

If we wish to visualize rolling Sharpe with `ggplot`, we can convert that xts object to a data frame and then pipe it, or we can start with one of our tidy `tibble` objects. The flow below starts with `xts` and converts to `tibble` with `tk_tbl()` so that we can get familiar with a new function.

```
rolling_sharpe_xts %>%
    tk_tbl(preserve_index = TRUE,
           rename_index = "date") %>%
    rename(rolling_sharpe = xts) %>%
    ggplot(aes(x = date,
               y = rolling_sharpe)) +
    geom_line(color = "cornflowerblue") +
    ggtitle("Rolling 24-Month Sharpe Ratio") +
    labs(y = "rolling sharpe ratio") +
```

```
scale_x_date(breaks = pretty_breaks(n = 8)) +
   theme(plot.title = element_text(hjust = 0.5))
```

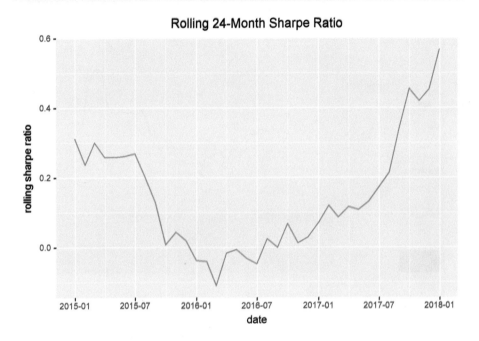

FIGURE 7.5: Rolling Sharpe ggplot

Figure 7.5 is showing the same data as Figure 7.4 but on a slightly more compressed scale. Would the scale variation lead us or an end user to think differently about this portfolio?

Those rolling charts allows us to see how our portfolio Sharpe Ratio decreased steadily into 2016, bottomed out, and then started to grind higher.

Let's take all this work and make it accessible via Shiny!

7.9 Shiny App Sharpe Ratio

The Sharpe Ratio Shiny app structure should feel familiar but have a quick look at the final app in Figure 7.6 and notice a few differences from our usual:

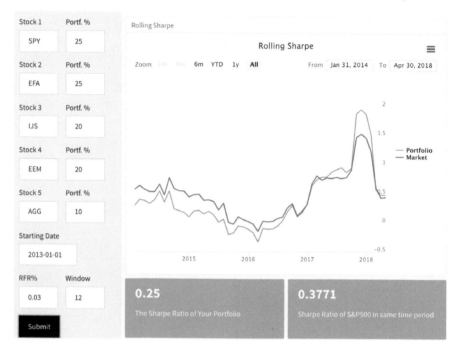

FIGURE 7.6: www.reproduciblefinance.com/shiny/sharpe-ratio/

Because the Sharpe Ratio is best understood by comparison we chart the rolling Sharpe Ratio of our portfolio alongside that of the S&P500, plus we have added two blue value boxes. That means we need to calculate the rolling and overall Sharpe for the S&P500 based on whatever starting date the user selects.

There are several calculations for this app and I divide them into market calculations and portfolio calculations.

In the chunks below, we run our market Sharpe Ratio equations, relying on the user-selected RFR, rolling window and starting date. The code flow runs through three reactives: `market_returns()`, which is used to find the `market_sharpe()` and the `market_rolling_sharpe()`.

First, we get the RFR, rolling window and market returns.

```
# market calculations

# Get the RFR from the end user
rfr <- eventReactive(input$go, {input$rfr/100})
```

```
# Get the rolling window from the end users
window <- eventReactive(input$go, {input$window})

# Calculate market returns based on starting date
market_returns <- eventReactive(input$go, {

    getSymbols("SPY", src = 'yahoo',
            from = input$date,
            auto.assign = TRUE,
            warnings = FALSE) %>%
    map(~Ad(get(.))) %>%
    reduce(merge) %>%
    `colnames<-`("SPY") %>%
    to.monthly(indexAt = "lastof",
            OHLC = FALSE) %>%
    Return.calculate(method = "log") %>%
    na.omit()
})
```

We now have a reactive object called `market_returns()`. Next, we calculate the overall and rolling market Sharpe Ratio of that object.

```
# Calculate market Sharpe Ratio
market_sharpe <- eventReactive(input$go, {

  SharpeRatio(market_returns(),
            Rf = rfr(),
            FUN = "StdDev")
})
```

```
# Calculate rolling market Sharpe Ratio
market_rolling_sharpe <- eventReactive(input$go, {

  rollapply(market_returns(),
            window(),
            function(x)
            SharpeRatio(x,
                    Rf = rfr(),
                    FUN = "StdDev")) %>%
  na.omit()
})
```

We will use two of those reactives in the main part of the app.

`market_sharpe()` appears in a blue value box and `market_rolling_sharpe()` appears on the `highcharter` chart.

Next, we calculate our *portfolio* Sharpe Ratio. The code flow is very similar to the previous, except we start by building portfolio returns.

```
# Run portfolio returns calculations

portfolio_returns <- eventReactive(input$go, {

  symbols <- c(input$stock1, input$stock2,
               input$stock3, input$stock4,
               input$stock5)

  validate(need(input$w1 + input$w2 +
                input$w3 + input$w4 +
                input$w5 == 100,
          "The portfolio weights must sum to 100%!"))

  w <- c(input$w1/100, input$w2/100,
         input$w3/100, input$w4/100,
         input$w5/100)

  getSymbols(symbols, src = 'yahoo', from = input$date,
             auto.assign = TRUE, warnings = FALSE) %>%
  map(~Ad(get(.))) %>%
  reduce(merge) %>%
  `colnames<-`(symbols) %>%
  to.monthly(indexAt = "lastof",
             OHLC = FALSE) %>%
  Return.calculate(method = "log") %>%
  na.omit() %>%
  Return.portfolio(weights = w)

})
```

We now have a reactive object called `portfolio_returns()`. Next, we calculate the overall and rolling market Sharpe Ratio of that object.

```
# Calculate portfolio Sharpe Ratio
portfolio_sharpe <- eventReactive(input$go, {

  validate(need(input$w1 + input$w2 + input$w3 +
                input$w4 + input$w5 == 100,
          "------"))
```

```
SharpeRatio(portfolio_returns(),
            Rf = rfr(),
            FUN = "StdDev")

})


# Calculate portfolio rolling Sharpe Ratio
portfolio_rolling_sharpe <- eventReactive(input$go, {

  rollapply(portfolio_returns(),
            window(),
            function(x) SharpeRatio(x,
                        Rf = rfr(),
                        FUN = "StdDev")) %>%
  na.omit()
})
```

We will use two objects from that code chunk in the main part of the app:
portfolio_sharpe() and portfolio_rolling_sharpe().

Note one crucial line in the above chunk: validate(need(input$w1
+ input$w2 + input$w3 + input$w4 + input$w5 == 100, ...)). This is
where we ensure that the weights sum to 100. If they do not, the user will see
an error message that reads "The portfolio weights must sum to 100%!"

Rolling Sharpe

The portfolio weights must sum to 100%!

FIGURE 7.7: Weights Error Message

We have not included that error message shown in Figure 7.7 in any of our
previous apps because we were introducing new concepts and did not want to
clutter the code. It is a good idea to include messages like this as guideposts
for our users.

Next, we pass our rolling market and rolling portfolio Sharpe Ratios to
renderHighchart(), and add a legend to the end of the code flow so that the
user can see which line is which.

```
# build one highchart that displays rolling Sharpe of both
# the portfolio and the market
renderHighchart({
```

```
validate(need(input$go, "Please choose your portfolio assets,
               weights, rfr, rolling window and start date
               and click submit."))

highchart(type = "stock") %>%
hc_title(text = "Rolling Sharpe") %>%
hc_add_series(portfolio_rolling_sharpe(),
              name = "Portfolio",
              color = "cornflowerblue") %>%
hc_add_series(market_rolling_sharpe(),
              name = "Market",
              color = "green") %>%
hc_navigator(enabled = FALSE) %>%
hc_scrollbar(enabled = FALSE) %>%
hc_exporting(enabled = TRUE) %>%
# Add a legend
hc_legend(enabled = TRUE,
          align = "right",
          verticalAlign = "middle",
          layout = "vertical")
})
```

Now we build and display the overall Sharpe Ratios of the portfolio and the market with two blue valueBox() aesthetics.

```
# value box for portfolio Sharpe Ratio
renderValueBox({
  valueBox(value = tags$p(round(portfolio_sharpe(), 4),
                          style = "font-size: 70%;"),
           color = "primary")
})
```

```
# value box for market Sharpe Ratio
renderValueBox({
  valueBox(value = tags$p(round(market_sharpe(), 4),
                          style = "font-size: 70%;"),
           color = "primary")
})
```

FIGURE 7.8: Sharpe Ratio Value Boxes

Figure 7.8 displays the value boxes. Maybe we love them and, more importantly, maybe our end users love them or despise them. The only way to know is to test and iterate, and that raises an important point about Shiny apps. Shiny involves experimentation because it depends on how end users experience the world. That is not a natural way to think about our work as a data scientist or quant. For example, as a data scientist or a quant, we might be perfectly satisfied to know that our code runs, calculates the correct rolling Sharpe Ratio and builds a nice data visualization. As a Shiny app builder, we must be concerned with whether an end user likes our work enough to interact with it, ideally more than once, and derive value from it. Considering the user experience is an exciting part of the Shiny challenge, indeed!

8

CAPM

By way of extraordinarily brief background, the Capital Asset Pricing Model (CAPM) is a model, created by William Sharpe, that estimates the return of an asset based on the return of the market and the asset's linear relationship to the return of the market. That linear relationship is the stock's beta coefficient. Beta can be thought of as the stock's sensitivity to the market, or its riskiness with respect to the market.

CAPM was introduced back in 1964, garnered a Nobel for its creator and, like many epoch-altering theories, has been widely used, updated, criticized, debunked, revived, re-debunked, etc. Fama and French have written that CAPM "is the centerpiece of MBA investment courses. Indeed, it is often the only asset pricing model taught in these courses... [u]nfortunately, the empirical record of the model is poor."[1]

Nevertheless, we will forge ahead with our analysis because calculating CAPM betas can serve as a nice template for more complex models. Plus, CAPM is still an iconic model. We will focus on one particular aspect of CAPM: beta. Beta, as we noted above, is the beta coefficient of an asset that results from regressing the returns of that asset on market returns. It captures the linear relationship between the asset and the market. For our purposes, it's a good vehicle for exploring a reproducible flow for modeling or regressing our portfolio returns on the market returns. Even if your team prefers more nuanced models, this workflow can serve as a good base.

8.1 CAPM and Market Returns

Our first step is to make a choice about which asset to use as a proxy for the market return and we will go with the SPY ETF, effectively treating the S&P500 as the market. That makes our calculations substantively uninteresting because (1) SPY is 25% of our portfolio and (2) we have chosen assets and

---

[1]Fama, Eugene F. and French, Kenneth R. "The Capital Asset Pricing Model: Theory and Evidence", The Journal of Economic Perspectives, Vol. 18, No. 3 (Summer, 2004), pp. 25-46.

a time period (2013 - 2017) in which correlations with SPY have been high. With those caveats in mind, feel free to choose a different asset for the market return and try to reproduce this work, or construct a different portfolio that does not include SPY.

We first import prices for SPY, calculate monthly returns and save the object as `market_returns_xts`.

```
market_returns_xts <-
    getSymbols("SPY",
               src = 'yahoo',
               from = "2012-12-31",
               to = "2017-12-31",
               auto.assign = TRUE,
               warnings = FALSE) %>%
    map(~Ad(get(.))) %>%
    reduce(merge) %>%
    `colnames<-`("SPY") %>%
    to.monthly(indexAt = "lastof",
               OHLC = FALSE) %>%
  Return.calculate(.,
                   method = "log") %>%
  na.omit()
```

We also want a `tibble` object of market returns for when we use the tidyverse.

```
market_returns_tidy <-
  market_returns_xts %>%
    tk_tbl(preserve_index = TRUE,
           rename_index = "date") %>%
    na.omit() %>%
    select(date, returns = SPY)
```

Since we will be regressing portfolio returns on market returns, let's ensure that the number of portfolio returns observations is equal to the number of market returns observations.

```
portfolio_returns_tq_rebalanced_monthly %>%
  mutate(market_returns = market_returns_tidy$returns) %>%
  head()
```

```
# A tibble: 6 x 3
  date          returns market_returns
  <date>          <dbl>          <dbl>
```

```
1 2013-01-31  0.0308           0.0499
2 2013-02-28 -0.000870         0.0127
3 2013-03-31  0.0187           0.0373
4 2013-04-30  0.0206           0.0190
5 2013-05-31 -0.00535          0.0233
6 2013-06-30 -0.0230          -0.0134
```

Note that if the number observations were not the same, `mutate()` would throw an error in the code chunk above.

8.2 Calculating CAPM Beta

Portfolio beta is equal to the covariance of the portfolio returns and market returns, divided by the variance of market returns. Here is the equation:

$$\beta_{portfolio} = cov(R_p, R_m)/\sigma_m$$

We calculate covariance of portfolio and market returns with `cov()`, and the variance of market returns with `var()`.

Our portfolio beta is equal to:

```
cov(portfolio_returns_xts_rebalanced_monthly,
    market_returns_tidy$returns)/
  var(market_returns_tidy$returns)
```

```
          [,1]
returns 0.8917
```

That beta is near to 1 and it is not a surprise since SPY is a big part of this portfolio.

We can also calculate portfolio beta by finding the beta of each of our assets and then multiplying by asset weights. That is, another equation for portfolio beta is the weighted sum of the asset betas:

$$\beta_{portfolio} = \sum_{i=1}^{n} W_i \, \beta_i$$

We first find the beta for each of our assets and this affords an opportunity to introduce a code flow for regression analysis.

We need to regress each of our individual asset returns on the market return and use the `lm()` function for that purpose. We could do that for asset 1 with

lm(asset_return_1 ~ market_returns_tidy$returns), and then again for
asset 2 with lm(asset_return_2 ~ market_returns_tidy$returns) etc. for
all 5 of our assets. But if we had a 50-asset portfolio, that would be impractical.
Instead we write a code flow and use map() to regress each of our asset returns
on market returns with one call.

We will start with our asset_returns_long tidy data frame and will then
run nest(-asset).

```
beta_assets <-
  asset_returns_long %>%
  nest(-asset)

beta_assets
```

```
# A tibble: 5 x 2
  asset data
  <chr> <list>
1 SPY   <tibble [60 x 2]>
2 EFA   <tibble [60 x 2]>
3 IJS   <tibble [60 x 2]>
4 EEM   <tibble [60 x 2]>
5 AGG   <tibble [60 x 2]>
```

That nest(-asset) changed our data frame so that there are two columns:
one called asset that holds the asset name and one called data that holds a
list of returns for each asset. We have 'nested' a list of returns within a column.

Now we use map() to apply the lm() function to each of those nested lists and
store the results in a new column called model.

```
beta_assets <-
  asset_returns_long %>%
  nest(-asset) %>%
  mutate(model =
           map(data, ~
                 lm(returns ~ market_returns_tidy$returns,
                      data = .)))

beta_assets
```

```
# A tibble: 5 x 3
  asset data              model
  <chr> <list>            <list>
1 SPY   <tibble [60 x 2]> <S3: lm>
```

```
2 EFA    <tibble [60 x 2]> <S3: lm>
3 IJS    <tibble [60 x 2]> <S3: lm>
4 EEM    <tibble [60 x 2]> <S3: lm>
5 AGG    <tibble [60 x 2]> <S3: lm>
```

We now have 3 columns: `asset` which we had before, `data` which we had before, and `model` which we just added. The `model` column holds the results of the regression `lm(returns ~ market_returns_tidy$returns, data = .)` that we ran for each of our assets. Those results are a beta and an intercept for each of our assets but stored as an S3 object, not ideal for presentation.

We can clean up the results with the `tidy()` function from the `broom` package. We want to apply that function to our model column and will use the `mutate()` and `map()` combination again.

```
beta_assets <-
  asset_returns_long %>%
  nest(-asset) %>%
  mutate(model =
           map(data, ~
                 lm(returns ~ market_returns_tidy$returns,
                    data = .))) %>%
  mutate(model = map(model, tidy))


beta_assets
```

```
# A tibble: 5 x 3
  asset data                model
  <chr> <list>              <list>
1 SPY   <tibble [60 x 2]> <data.frame [2 x 5]>
2 EFA   <tibble [60 x 2]> <data.frame [2 x 5]>
3 IJS   <tibble [60 x 2]> <data.frame [2 x 5]>
4 EEM   <tibble [60 x 2]> <data.frame [2 x 5]>
5 AGG   <tibble [60 x 2]> <data.frame [2 x 5]>
```

We are getting close now with a list of `tibbles` in the `model` column.

For readability, we `unnest()` that `model` column.

```
beta_assets <-
  asset_returns_long %>%
  nest(-asset) %>%
  mutate(model =
           map(data, ~
                 lm(returns ~ market_returns_tidy$returns,
                    data = .))) %>%
```

```
mutate(model = map(model, tidy)) %>%
unnest(model) %>%
mutate_if(is.numeric, funs(round(., 4)))
```

```
beta_assets
```

```
# A tibble: 10 x 6
   asset term       estimate std.error statistic p.value
   <chr> <chr>         <dbl>     <dbl>     <dbl>   <dbl>
 1 SPY   (Interc~  0.        0.        0.         1.00
 2 SPY   market_~  1.00      0.        3.12e+16   0.
 3 EFA   (Interc~ -0.00550   0.00290  -1.89e+ 0   0.0632
 4 EFA   market_~  0.941     0.0974    9.66e+ 0   0.
 5 IJS   (Interc~ -0.00170   0.00360  -4.73e- 1   0.638
 6 IJS   market_~  1.12      0.122     9.19e+ 0   0.
 7 EEM   (Interc~ -0.00840   0.00480  -1.74e+ 0   0.0863
 8 EEM   market_~  0.919     0.162     5.67e+ 0   0.
 9 AGG   (Interc~  0.00170   0.00120   1.40e+ 0   0.166
10 AGG   market_~ -0.0109    0.0417   -2.61e- 1   0.795
```

Now that looks more readable and presentable. We will go one step further
and delete the intercept results since we are isolating the betas.

```
beta_assets <-
  asset_returns_long %>%
  nest(-asset) %>%
  mutate(model =
           map(data, ~
                lm(returns ~ market_returns_tidy$returns,
                   data = .))) %>%
  unnest(model %>% map(tidy)) %>%
  filter(term != "(Intercept)") %>%
  select(-term)
```

```
beta_assets
```

```
# A tibble: 5 x 5
  asset estimate std.error statistic  p.value
  <chr>    <dbl>     <dbl>     <dbl>    <dbl>
1 SPY    1.00     3.21e-17  3.12e+16 0.
2 EFA    0.941    9.74e- 2  9.66e+ 0 1.08e-13
3 IJS    1.12     1.22e- 1  9.19e+ 0 6.47e-13
4 EEM    0.919    1.62e- 1  5.67e+ 0 4.81e- 7
5 AGG   -0.0109   4.17e- 2 -2.61e- 1 7.95e- 1
```

A quick sanity check on those asset betas should reveal that SPY has beta of 1 with itself.

```
beta_assets %>%
  select(asset, estimate) %>%
  filter(asset == "SPY")
```

```
# A tibble: 1 x 2
  asset estimate
  <chr>    <dbl>
1 SPY        1.
```

Now we use a weighted combination of these assets to calculate total portfolio beta.

We use our original weights assigning vector w and multiply by the betas of each asset.

```
beta_byhand <-
  w[1] * beta_assets$estimate[1] +
  w[2] * beta_assets$estimate[2] +
  w[3] * beta_assets$estimate[3] +
  w[4] * beta_assets$estimate[4] +
  w[5] * beta_assets$estimate[5]

beta_byhand
```

```
[1] 0.8917
```

That beta is the same as we calculated before. We have confirmed that the the covariance of portfolio returns and market returns divided by the variance of market returns is equal to the weighted beta estimate for each asset. Now let's confirm consistent results with built-in functions.

8.3 Calculating CAPM Beta in the xts world

We start in the xts world and use the built-in CAPM.beta() function from PerformanceAnalytics. That function takes two arguments: the returns for the portfolio whose beta we wish to calculate, and the market returns.

```
beta_builtin_xts <-
  CAPM.beta(portfolio_returns_xts_rebalanced_monthly,
            market_returns_xts)
beta_builtin_xts
```

```
[1] 0.8917
```

8.4   Calculating CAPM Beta in the tidyverse

Finding and displaying portfolio beta in the tidyverse requires regressing our
portfolio returns object on the market returns object, and then cleaning up
the results with broom.

```
beta_dplyr_byhand <-
  portfolio_returns_tq_rebalanced_monthly %>%
  do(model =
       lm(returns ~ market_returns_tidy$returns,
          data = .)) %>%
  tidy(model) %>%
  mutate(term = c("alpha", "beta")) %>%
  select(estimate)

beta_dplyr_byhand$estimate[2]
```

```
[1] 0.8917
```

8.5   Calculating CAPM Beta in the tidyquant world

Finally, we use tq_performance() to apply CAPM.beta() to our returns object.

```
beta_builtin_tq <-
  portfolio_returns_tq_rebalanced_monthly %>%
  mutate(market_return =
           market_returns_tidy$returns) %>%
```

```
na.omit() %>%
tq_performance(Ra = returns,
               Rb = market_return,
               performance_fun = CAPM.beta) %>%
`colnames<-`("beta_tq")
```

Let's take a quick look at our four beta calculations.

```
beta_builtin_tq %>%
  mutate(dplyr_beta = beta_dplyr_byhand$estimate[2],
         byhand_beta = beta_byhand,
         xts_beta = coredata(beta_builtin_xts)) %>%
  round(3)
```

```
# A tibble: 1 x 4
  beta_tq dplyr_beta byhand_beta xts_beta
    <dbl>      <dbl>       <dbl>    <dbl>
1   0.892      0.892       0.892    0.892
```

Consistent results and a beta near 1, as expected since our portfolio has a 25% allocation to the S&P500. We used regression, we used a by-hand calculation, we used built-in functions and hopefully we now have a good grasp on what people mean when they say CAPM beta.

8.6 Visualizing CAPM with ggplot

Let's start our CAPM visualization with a scatter plot of market returns on the x-axis and portfolio returns on the y-axis.

```
portfolio_returns_tq_rebalanced_monthly %>%
  mutate(market_returns =
            market_returns_tidy$returns) %>%
  ggplot(aes(x = market_returns,
             y = returns)) +
  geom_point(color = "cornflowerblue") +
  ylab("portfolio returns") +
  xlab("market returns")
```

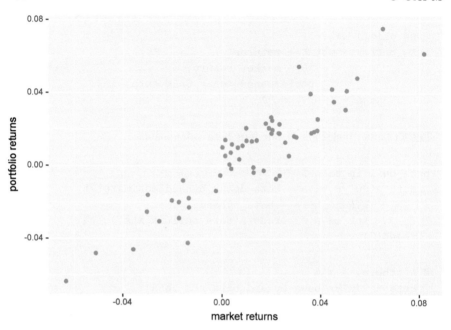

FIGURE 8.1: Scatter Portfolio v. Market

Our beta calculation communicated a strong linear relationship, and Figure 8.1 is communicating the same.

We can add a simple regression line with `geom_smooth(method = "lm"`, `se = FALSE,...)`. Have a look at Figure 8.2 to see how the line appears.

```
portfolio_returns_tq_rebalanced_monthly %>%
  mutate(market_returns =
           market_returns_tidy$returns) %>%
  ggplot(aes(x = market_returns,
             y = returns)) +
  geom_point(color = "cornflowerblue") +
  geom_smooth(method = "lm",
              se = FALSE,
              color = "green") +
  ylab("portfolio returns") +
  xlab("market returns")
```

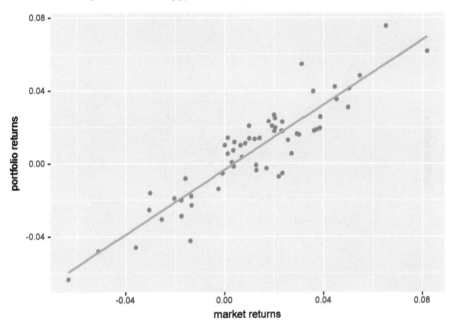

FIGURE 8.2: Scatter with Regression Line from ggplot

The green line in Figure 8.2 was produced by the call to `geom_smooth(method = 'lm')`. Under the hood, `ggplot()` fit a linear model of the relationship between market returns and portfolio returns. The slope of that green line equals the CAPM beta that we calculated earlier. To confirm that, we can add a line to the scatter that has a slope equal to our beta calculation and a y-intercept equal to what I labeled as alpha in the `beta_dplyr_byhand` object.

To add the line, we call `geom_abline(...)`.

```
portfolio_returns_tq_rebalanced_monthly %>%
  mutate(market_returns = market_returns_tidy$returns) %>%
  ggplot(aes(x = market_returns, y = returns)) +
  geom_point(color = "cornflowerblue") +
  geom_abline(aes(
    intercept = beta_dplyr_byhand$estimate[1],
    slope = beta_dplyr_byhand$estimate[2]),
              color = "purple") +
  ylab("portfolio returns") +
  xlab("market returns")
```

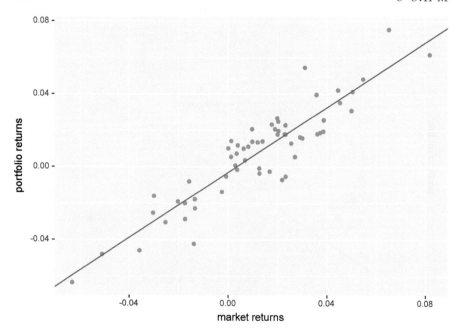

FIGURE 8.3: Scatter with Regression Line from Beta Estimate

The purple line in Figure 8.3 was built using our own beta calculations from before.

We can plot both lines simultaneously to confirm that they are the same - they should be right on top of each other but the purple line, our manual `abline`, extends into infinity so we should see it start where the green line ends.

```
portfolio_returns_tq_rebalanced_monthly %>%
  mutate(market_returns =
           market_returns_tidy$returns) %>%
  ggplot(aes(x = market_returns,
             y = returns)) +
  geom_point(color = "cornflowerblue") +
  geom_abline(
    aes(intercept =
          beta_dplyr_byhand$estimate[1],
        slope = beta_dplyr_byhand$estimate[2]),
            color = "purple") +
  geom_smooth(method = "lm",
              se = FALSE,
              color = "green") +
```

```
ylab("portfolio returns") +
xlab("market returns")
```

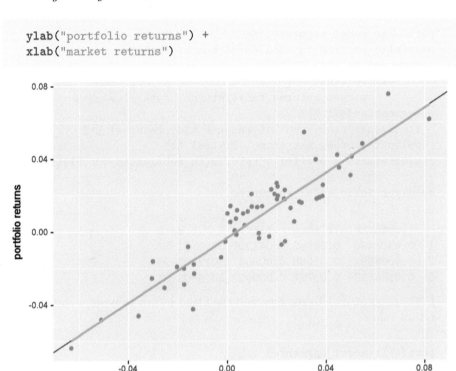

FIGURE 8.4: Scatter with Both Regression Lines

Figure 8.4 seems to visually confirm (or strongly support) that the fitted line calculated by `ggplot()` and `geom_smooth()` has a slope equal to the beta we calculated by-hand. Why did we go through this exercise? CAPM beta is a bit jargony and it's useful to consider how jargon reduces to data science concepts. We might need to explain it someday and what better way than with a data visualization.

8.7 Augmenting Our Data

Before concluding our analysis of CAPM beta, we will explore the `augment()` function from `broom`.

The code chunk below will start with model results from `lm(returns ~ market_returns_tidy$returns...)` and then `augment()` the original data set with predicted values, stored in a new column called `.fitted`.

```
portfolio_model_augmented <-
portfolio_returns_tq_rebalanced_monthly %>%
  do(model =
       lm(returns ~
            market_returns_tidy$returns, data = .)) %>%
  augment(model) %>%
  rename(mkt_rtns = market_returns_tidy.returns) %>%
  select(returns, mkt_rtns, .fitted) %>%
  mutate(date = portfolio_returns_tq_rebalanced_monthly$date)

head(portfolio_model_augmented, 3)


      returns mkt_rtns   .fitted        date
1   0.0308488  0.04992 0.041308 2013-01-31
2  -0.0008697  0.01268 0.008097 2013-02-28
3   0.0186623  0.03727 0.030023 2013-03-31
```

Let's use `ggplot()` to see how well the fitted return values match the actual return values.

```
portfolio_model_augmented %>%
  select(date, returns, .fitted) %>%
  gather(type, data, -date) %>%
  ggplot(aes(x = date, y = data, color = type)) +
  geom_line() +
  xlab("date")
```

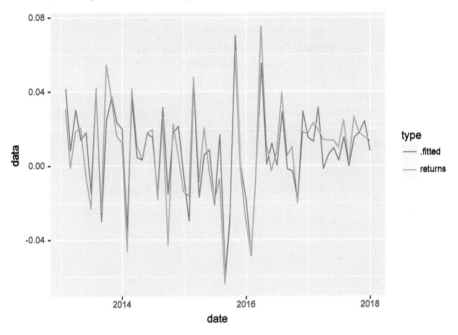

FIGURE 8.5: Actual versus Fitted Returns

Figure 8.5 indicates that our fitted values track the actual values quite well. Now let's get to some more complicated `highcharter` work.

8.8 Visualizing CAPM with highcharter

One benefit of `augment()` is that it allows us to create an interesting `highcharter` visualization that replicates our scatter plot with regression `ggplot` in Figure 8.2.

First, let's build the base scatter plot of portfolio returns, which are housed in `portfolio_model_augmented$returns`, against market returns, which are housed in `portfolio_model_augmented$mkt_rtns`.

```
highchart() %>%
  hc_title(text = "Portfolio v. Market Returns Scatter") %>%
  hc_add_series(portfolio_model_augmented,
               type = "scatter",
```

```
                  color = "cornflowerblue",
                  hcaes(x = round(mkt_rtns, 4),
                       y = round(returns, 4)),
                  name = "Returns") %>%
  hc_xAxis(title = list(text = "Market Returns")) %>%
  hc_yAxis(title = list(text = "Portfolio Returns")) %>%
  hc_add_theme(hc_theme_flat()) %>%
  hc_exporting(enabled = TRUE)
```

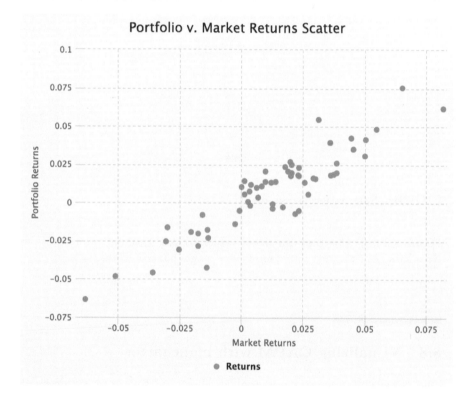

FIGURE 8.6: CAPM Scatter highcharter

The scatter plot in Figure 8.6 looks good but let's add a functionality so that when a user hovers on a point, the date is displayed.

First, we need to supply the date observations, so we will add a date variable with `hc_add_series_scatter(..., date = portfolio_returns_tq_rebalanced_monthly$date)`.

Then we want the tool tip to pick up and display that variable. That is done with `hc_tooltip(formatter = JS("function(){return ('port return:`

' + this.y + '
 mkt return: ' + this.x + '
 date: ' + this.point.date)}")).

We are creating a custom tool tip function to pick up the date.

```
highchart() %>%
  hc_title(text = "Scatter Plot with Date") %>%
  hc_add_series(portfolio_model_augmented,
                type = "scatter",
                color = "cornflowerblue",
                hcaes(x = round(mkt_rtns, 4),
                      y = round(returns, 4),
                      date = date),
                name = "Returns") %>%
  hc_xAxis(title = list(text = "Market Returns")) %>%
  hc_yAxis(title = list(text = "Portfolio Returns")) %>%
  hc_tooltip(formatter = JS("function(){
    return ('port return: ' + this.y +
    ' <br> mkt return: ' + this.x +
    ' <br> date: ' + this.point.date)}")) %>%
  hc_add_theme(hc_theme_flat()) %>%
  hc_exporting(enabled = TRUE)
```

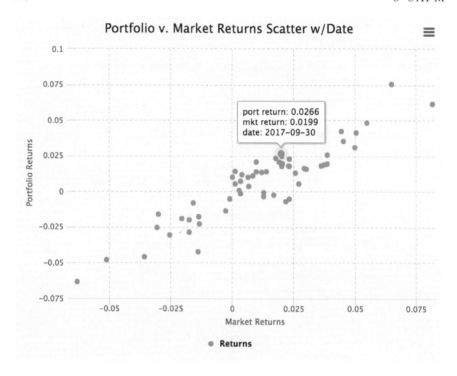

FIGURE 8.7: CAPM highcharter scatter with Date

Figure 8.7 displays how the chart behaves when a user hovers on a point. Note that the date is displayed so we can see *when* any interesting observations occurred.

Finally, let's add the regression line.

To do that, we need to supply x- and y-coordinates to `highcharter` and specify that we want to add a line instead of more scatter points. We have the x- and y-coordinates for our fitted regression line because we added them with the `augment()` function. The x's are the market returns and the y's are the fitted values. We add this element to our code flow with `hc_add_series(portfolio_model_augmented, type = "line", hcaes(x = mkt_rtns, y = .fitted))`.

```
highchart() %>%
  hc_title(text = "Scatter with Regression Line") %>%
  hc_add_series(portfolio_model_augmented,
                type = "scatter",
                color = "cornflowerblue",
```

```
            hcaes(x = round(mkt_rtns, 4),
                  y = round(returns, 4),
                  date = date),
            name = "Returns") %>%
  hc_add_series(portfolio_model_augmented,
                type = "line",
                hcaes(x = mkt_rtns, y = .fitted),
                name = "CAPM Beta = Regression Slope") %>%
  hc_xAxis(title = list(text = "Market Returns")) %>%
  hc_yAxis(title = list(text = "Portfolio Returns")) %>%
  hc_tooltip(formatter = JS("function(){
    return ('port return: ' + this.y + ' <br> mkt return: '
    + this.x + ' <br> date: ' + this.point.date)}")) %>%
  hc_add_theme(hc_theme_flat()) %>%
  hc_exporting(enabled = TRUE)
```

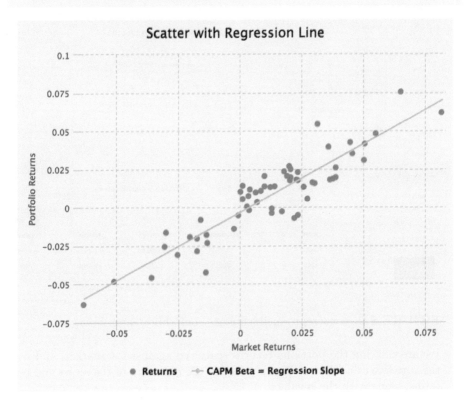

FIGURE 8.8: CAPM highcharter scatter with LM

Figure 8.8 chart depicts the same substance as we displayed in Figure 8.2 with ggplot() but this one is more interactive. In a Shiny app (or HTML report), the user can hover on the graph and see the date.

Let's head to Shiny and let an end user run custom CAPM calculations..

8.9   Shiny App CAPM

Our CAPM Shiny app allows a user to build a portfolio and calculate its market beta, while also plotting portfolio returns against market returns.

Have a look at the final app in Figure 8.9 and note that the visualizations are different from what we have built previously, but the inputs are the same.

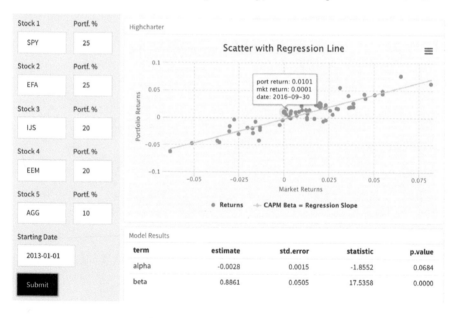

FIGURE 8.9: www.reproduciblefinance.com/shiny/capm-beta/

We are showing the portfolio returns scattered against the market, and with the regression line drawn. We also display the estimates of the alpha and beta terms, along with the p-values.

Our input sidebar and portfolio returns calculations are our standard.

Similar to our Sharpe Ratio app, we calculate the market return but we convert to a `tibble` for use as our independent variable in the call to `lm()`.

```
portfolio_returns_tq_rebalanced_monthly <-
  eventReactive(input$go, {
  prices <- prices()
  w <- c(input$w1/100, input$w2/100, input$w3/100,
         input$w4/100, input$w5/100)

  # Convert returns to a tibble

  portfolio_returns_tq_rebalanced_monthly <-
      prices %>%
      to.monthly(indexAt = "lastof", OHLC = FALSE) %>%
      tk_tbl(preserve_index = TRUE, rename_index = "date") %>%
      gather(asset, returns, -date) %>%
      group_by(asset) %>%
      mutate(returns = (log(returns) - log(lag(returns)))) %>%
      na.omit() %>%
      tq_portfolio(assets_col  = asset,
               returns_col = returns,
               weights     = w,
               col_rename  = "returns",
               rebalance_on = "months")
})

# Calculate market return and convert to a tibble.
market_return <- eventReactive(input$go, {
  market_return <-
    getSymbols("SPY",
               src = 'yahoo',
               from = input$date,
             auto.assign = TRUE,
             warnings = FALSE) %>%
    map(~Ad(get(.))) %>%
    reduce(merge) %>%
    `colnames<-`("SPY") %>%
    to.monthly(indexAt = "lastof",
               OHLC = FALSE) %>%
    tk_tbl(preserve_index = TRUE,
           rename_index = "date") %>%
    mutate(returns = (log(SPY) - log(lag(SPY)))) %>%
    na.omit() %>%
    select(date, returns)
})
```

We now have two reactives, `portfolio_returns_tq_rebalanced_monthly()` and `market_return()` to run our regression and calculate beta.

```
beta_dplyr_byhand <- eventReactive(input$go, {

  portfolio_returns_tq_rebalanced_monthly <-
    portfolio_returns_tq_rebalanced_monthly()

  market_return <- market_return()

beta_dplyr_byhand <-
  portfolio_returns_tq_rebalanced_monthly %>%
  do(model = lm(returns ~ market_return$returns, data = .))

})
```

Next we create the augmented model to facilitate our `highcharter` visualization.

```
portfolio_model_augmented <- eventReactive(input$go, {

  portfolio_returns_tq_rebalanced_monthly <-
    portfolio_returns_tq_rebalanced_monthly()

  beta_dplyr_byhand() %>%
  augment(model) %>%
  rename(mkt_rtns = market_return.returns) %>%
  select(returns, mkt_rtns, .fitted) %>%
  mutate(date = portfolio_returns_tq_rebalanced_monthly$date)

})
```

And, finally, call `renderHighchart()` to create the scatter plot and regression line.

```
renderHighchart({

portfolio_model_augmented <- portfolio_model_augmented()

highchart() %>%
  hc_title(text = "Scatter with Regression Line") %>%
  hc_add_series(portfolio_model_augmented,
                type = "scatter",
```

```
                color = "cornflowerblue",
                hcaes(x = round(mkt_rtns, 4),
                      y = round(returns, 4),
                      date = date),
                name = "Returns") %>%
  hc_add_series(portfolio_model_augmented,
                type = "line",
                enableMouseTracking = FALSE,
                hcaes(x = mkt_rtns, y = .fitted),
                name = "CAPM Beta = Slope of Line") %>%
  hc_xAxis(title = list(text = "Market Returns")) %>%
  hc_yAxis(title = list(text = "Portfolio Returns")) %>%
  hc_tooltip(formatter = JS("function(){
    return ('port return: ' + this.y +
    ' <br> mkt return: ' + this.x +
    ' <br> date: ' + this.point.date)}"))%>%
  hc_add_theme(hc_theme_flat()) %>%
  hc_exporting(enabled = TRUE)

})
```

We add one new feature with `renderTable()` to show model results in a nice and neat table.

```
renderTable({
  beta_dplyr_byhand() %>%
  tidy(model) %>%
  mutate(term = c("alpha", "beta"))
}, digits = 4)
```

This app displays our CAPM beta calculations and from a general perspective, it's a good template for displaying the results of any simple linear regression. If we wished to regress our portfolio returns on a variable other than market returns, we would substitute the new variable into the template with a new `eventReactive()` but would keep most of the work unchanged. That can be useful when we show this app to an end user for general feedback and receive a request to replicate it with different variables for the regression. Alternatively, we could wire a Shiny app to let the end user choose from a selection of independent variables and test different regressions - consider how we might do that with a `selectInput()` that takes a ticker and grabs prices from a database or a source like Yahoo! Finance.

9

Fama-French Factor Model

We now move beyond CAPM's simple linear regression by adding variables to the equation and explore the Fama-French (FF) factor model of equity returns and risk.[1]

FF extends CAPM by regressing portfolio returns on several variables, in addition to market returns. From a general data science point of view, FF extends CAPM's simple linear regression, where we had one independent variable, to a multiple linear regression, where we have numerous independent variables.

We will look at an FF 3-factor model (there are also a number of 5-factor models), which tests the explanatory power of (1) market returns (same as CAPM), (2) firm size (small versus big) and (3) firm value. The firm value factor is labeled as HML in FF, which stands for high-minus-low and refers to a firm's book-to-market ratio. When we regress portfolio returns on the HML factor, we are investigating how much of the returns are the result of including stocks with a high book-to-market ratio (sometimes called the value premium, because high book-to-market stocks are called value stocks).

We have three goals for this chapter:

First, explore the surface of the FF model, because it has become quite important to portfolio theory.

Second, see how our work translates from simple linear regression to multiple linear regression, since most of our work in practice will be of the multiple variety.

Third, the most challenging part of this chapter, import data from a new source and wrangle it for use with our core data objects.

Since we will spend so much time on data wrangling, we will not look at different coding paradigms in this chapter. Our code will be in just the tidyverse and tidyquant realms.

We will see that wrangling the data is conceptually easy to understand but practically time-consuming to implement. However, mashing together data from disparate sources is a necessary skill for anyone in industry that has data

---

[1] Fama, Eugene and French, Kenneth. "Common risk factors in the returns on stocks and bonds." Journal of Financial Economics Volume 33, Issue 1, February 1993, Pages 3-56.

streams from different vendors and wants to get creative about how to use them. Once the data are wrangled, fitting the model is not time consuming.

9.1 Importing and Wrangling Fama-French Data

Our first task is to get the FF data and, fortunately, it is available at the FF website:

http://mba.tuck.dartmouth.edu/pages/faculty/ken.french/data_ library.html

We will document each step for importing and cleaning this data, to an extent that might be overkill. This may seem frustratingly tedious now, but it will be a time saver later if we needed to update this model or extend to the 5-factor case.

The data are packaged as zip files so we will need to do a bit more than call `read_csv()`.

First, we use the `tempfile()` function from base R to create a variable called `temp`.

```
temp <- tempfile()
```

R has created a temporary file called `temp` that will be cleaned up when we exit this session.

The URL for the 3-factor zip is http://mba.tuck.dartmouth.edu/pages/ faculty/ken.french/ftp/Global_3_Factors_CSV.zip. We want to pass that to `download.file()` and store the result in `temp`.

First, though, we will break that string into three pieces: `base`, `factor` and `format`. We will then `paste()` those together and save the string as `full_url`.

```
# Split the url into pieces
base <-
"http://mba.tuck.dartmouth.edu/pages/faculty/ken.french/ftp/"
factor <-
  "Global_3_Factors"
format<-
  "_CSV.zip"

# Paste the pieces together to form the full url
```

```
full_url <-
  paste(base,
        factor,
        format,
        sep ="")
```

Now we pass `full_url` to `download.file()`.

```
download.file(
full_url,
temp,
quiet = TRUE)
```

We want to read the csv file using `read_csv()` but first we need to unzip that data with the `unz()` function.

```
Global_3_Factors <-
  read_csv(unz(temp,
               "Global_3_Factors.csv"))
```

```
head(Global_3_Factors, 3)
```

```
# A tibble: 3 x 1
  `This file was created using the 201804 Bloomberg d~
  <chr>
1 Missing data are indicated by -99.99.
2 <NA>
3 199007
```

We have imported the dataset but we do not see any factors, just a column with weirdly formatted dates.

When this occurs, it *often* can be fixed by skipping a certain number of rows that contain metadata. Have a look at what happens if we skip 6 rows.

```
Global_3_Factors <-
  read_csv(unz(temp,
               "Global_3_Factors.csv"),
    skip = 6)
```

```
head(Global_3_Factors, 3)
```

```
# A tibble: 3 x 5
  X1       `Mkt-RF` SMB    HML    RF
```

```
    <chr>  <chr>     <chr> <chr> <chr>
1 199007 0.86       0.77  -0.25 0.68
2 199008 -10.82    -1.60  0.60  0.66
3 199009 -11.98     1.23  0.81  0.60
```

This is what were were expecting, 5 columns: X1 holds the weirdly formatted dates, Mkt-Rf holds the market returns above the risk-free rate, SMB hold the size factor, HML holds the value factor, and RF holds the risk-free rate.

However, the data have been coerced to a character format - look at the class of each column.

```
map(Global_3_Factors, class)
```

```
$X1
[1] "character"

$`Mkt-RF`
[1] "character"

$SMB
[1] "character"

$HML
[1] "character"

$RF
[1] "character"
```

We have two options for coercing those columns to the right format. We can do so upon import, by supplying the argument `col_types = cols(col_name = col_double(),...` for each numeric column.

```
Global_3_Factors <-
  read_csv(unz(temp,
               "Global_3_Factors.csv"),
           skip = 6,
           col_types = cols(
             `Mkt-RF` = col_double(),
             SMB = col_double(),
             HML = col_double(),
             RF = col_double()))
```

```
head(Global_3_Factors, 3)
```

```
# A tibble: 3 x 5
```

```
    X1      `Mkt-RF`   SMB     HML     RF
    <chr>       <dbl>  <dbl>   <dbl>  <dbl>
1 199007      0.860   0.770  -0.250  0.680
2 199008     -10.8   -1.60    0.600  0.660
3 199009     -12.0    1.23    0.810  0.600
```

That works well but it's specific to the FF 3-factor set with those specific column names. If we imported a different FF factor set, we would need to specify different column names.

The code chunk below converts the columns to numeric after import but is more general. It can be applied to other FF factor collections.

First, we rename the X1 column to `date`, and then use the `dplyr` verb `mutate_at(vars(-date), as.numeric)` to change our column formats to numeric. The `vars()` function operates like the `select()` function in that we can tell it to operate on all columns except the `date` column by putting a negative sign in front of `date`.

```
Global_3_Factors <-
  read_csv(unz(temp,
               "Global_3_Factors.csv"),
           skip = 6) %>%
  rename(date = X1) %>%
  mutate_at(vars(-date), as.numeric)

head(Global_3_Factors, 3)
```

```
# A tibble: 3 x 5
    date    `Mkt-RF`   SMB     HML     RF
    <chr>       <dbl>  <dbl>   <dbl>  <dbl>
1 199007      0.860   0.770  -0.250  0.680
2 199008     -10.8   -1.60    0.600  0.660
3 199009     -12.0    1.23    0.810  0.600
```

We now have numeric data for our factors and the date column has a better label, but the wrong format.

We can use the `lubridate` package to parse that date string into a nicer date format. We will use the `parse_date_time()` function, and call the `ymd()` function to make sure the end result is in a date format. Again, when working with data from a new source, the date and, indeed, any column can come in so many formats.

```
Global_3_Factors <-
  read_csv(unz(temp, "Global_3_Factors.csv"),
           skip = 6) %>%
```

```
rename(date = X1) %>%
mutate_at(vars(-date), as.numeric) %>%
mutate(date =
          ymd(parse_date_time(date, "%Y%m")))
```

```
head(Global_3_Factors, 3)
```

```
# A tibble: 3 x 5
  date        `Mkt-RF`    SMB    HML    RF
  <date>         <dbl>  <dbl>  <dbl> <dbl>
1 1990-07-01     0.860  0.770 -0.250 0.680
2 1990-08-01   -10.8   -1.60   0.600 0.660
3 1990-09-01   -12.0    1.23   0.810 0.600
```

The date format looks good now and that matters because we want to trim the factor dates to match our portfolio dates.

For our next challenge, notice that FF uses the first of the month and our portfolio object uses the last of the month. Fortunately, `lubridate` contains the `rollback()` function. This will roll monthly dates back to the last day of the previous month. The first date in our FF data is "1990-07-01". Let's roll it back.

```
Global_3_Factors %>%
  select(date) %>%
  mutate(date = lubridate::rollback(date)) %>%
  head(1)
```

```
# A tibble: 1 x 1
  date
  <date>
1 1990-06-30
```

If we want to reset our dates to the last of the same month, we need to add one first, then rollback. That might not be necessary, perhaps we are fine with rolling July back to June 30, for example. But if we want to keep July 1 as the "July" observation, this is how we accomplish it.

```
Global_3_Factors %>%
  select(date) %>%
  mutate(date = lubridate::rollback(date + months(1))) %>%
  head(1)
```

```
# A tibble: 1 x 1
  date
```

```
   <date>
1 1990-07-31
```

Our final code flow for wrangling the FF object is:

```
Global_3_Factors <-
  read_csv(unz(temp, "Global_3_Factors.csv"),
          skip = 6) %>%
  rename(date = X1) %>%
  mutate_at(vars(-date), as.numeric) %>%
  mutate(date =
          ymd(parse_date_time(date, "%Y%m"))) %>%
  mutate(date = rollback(date + months(1)))

  head(Global_3_Factors, 3)
```

```
# A tibble: 3 x 5
  date        `Mkt-RF`    SMB     HML     RF
  <date>         <dbl>  <dbl>   <dbl>  <dbl>
1 1990-07-31     0.860  0.770  -0.250  0.680
2 1990-08-31   -10.8   -1.60    0.600  0.660
3 1990-09-30   -12.0    1.23    0.810  0.600
```

There are other ways we could have gotten around this issue. Most notably, way back in the *Returns* section, we could have indexed our portfolio returns to `indexAt = firstof`,[2] but it was a good chance to introduce the `rollback()` function and we will not always have the option change our returns object. Sometimes two data sets are thrown at us and we have to wrangle them from there.

All that work enables us to merge these data objects together with `left_join(...by = "date")`. This will eliminate any rows with non-matching dates. We also convert the FF data to decimal format and create a new column called `R_excess` to hold our portfolio returns above the risk-free rate.

```
ff_portfolio_returns <-
  portfolio_returns_tq_rebalanced_monthly %>%
  left_join(Global_3_Factors, by = "date")  %>%
  mutate(MKT_RF = `Mkt-RF`/100,
         SMB = SMB/100,
         HML = HML/100,
         RF = RF/100,
         R_excess = round(returns - RF, 4)) %>%
```

---

[2]See this code flow: www.reproduciblefinance.com/code/ff-3-alt-wrangle

```
select(-returns, -RF)
```

```
head(ff_portfolio_returns, 3)
```

```
# A tibble: 3 x 6
   date        `Mkt-RF`     SMB      HML   MKT_RF
   <date>        <dbl>    <dbl>    <dbl>    <dbl>
1 2013-01-31     5.46   0.00140   0.0201   0.0546
2 2013-02-28    0.100   0.00330  -0.00780  0.00100
3 2013-03-31     2.29   0.00830  -0.0203   0.0229
# ... with 1 more variable: R_excess <dbl>
```

We now we have one object with our portfolio returns and FF factors and can proceed to the simplest part of our exercise: the modeling. It is simple because it follows the same code flow as CAPM, though we will also include the 95% confidence interval for our coefficients. We do that by setting `tidy(model, conf.int = T, conf.level = .95)`.

```
ff_dplyr_byhand <-
  ff_portfolio_returns %>%
  do(model =
       lm(R_excess ~ MKT_RF + SMB + HML,
              data = .)) %>%
  tidy(model, conf.int = T, conf.level = .95) %>%
  rename(beta = estimate)
```

```
ff_dplyr_byhand %>%
  mutate_if(is.numeric, funs(round(., 3))) %>%
  select(-statistic, -std.error)
```

```
          term    beta p.value conf.low conf.high
1 (Intercept)  -0.001   0.191   -0.004     0.001
2       MKT_RF   0.894   0.000    0.823     0.966
3          SMB   0.056   0.462   -0.095     0.207
4          HML   0.030   0.629   -0.092     0.151
```

9.2 Visualizing Fama-French with ggplot

Let's do something different with our visualization of the factor betas and add the confidence intervals to a scatter plot of coefficients. We will filter out the intercept term with `filter(term != "(Intercept)")` and add confidence intervals with `geom_errorbar(aes(ymin = conf.low, ymax = conf.high))`.

```
ff_dplyr_byhand %>%
  mutate_if(is.numeric, funs(round(., 3))) %>%
  filter(term != "(Intercept)") %>%
  ggplot(aes(x = term,
             y = beta,
             shape = term,
             color = term)) +
  geom_point() +
  geom_errorbar(aes(ymin = conf.low,
                    ymax = conf.high)) +
  labs(title = "FF 3-Factor Coefficients",
       subtitle = "balanced portfolio",
       x = "",
       y = "coefficient",
       caption = "data source: Fama-French website") +
  theme_minimal() +
  theme(plot.title = element_text(hjust = 0.5),
        plot.subtitle = element_text(hjust = 0.5),
        plot.caption = element_text(hjust = 0))
```

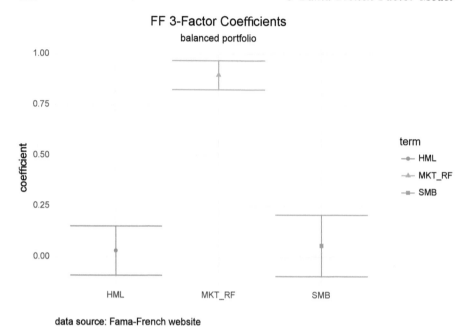

FIGURE 9.1: Fama-French factor betas

Figure 9.1 indicates that the market factor dominates this model and the other two factors contain zero in their confidence bands, which probably results from the fact that the market is actually a part of our portfolio. We will include this same visualization in our Shiny app so users can chart the betas of custom portfolios.

That completes our overall FF 3-factor model of this portfolio. The import and wrangling were the hardest challenge and that will often be the case. Running a model after the data cleaning and wrangling is the fun part.

9.3 Rolling Fama-French with the tidyverse and tibble-time

Before we head to Shiny, let's examine the rolling FF results and explore the model in different time periods. The following work is specific to FF, but can be applied generally to any multiple linear regression model that we wish to fit on a rolling basis.

We first define a rolling model with the `rollify()` function from `tibbletime`. Instead of wrapping an existing function, such as `kurtosis()` or `skewness()`, we will pass in our linear F3-factor model.

```
# Choose a 24-month rolling window
window <- 24
# define a rolling ff model with tibbletime
rolling_lm <-
  rollify(.f = function(R_excess, MKT_RF, SMB, HML) {
  lm(R_excess ~ MKT_RF + SMB + HML)
  }, window = window, unlist = FALSE)
```

Next, we pass columns from `ff_portfolio_returns` to the rolling function model.

```
rolling_ff_betas <-
  ff_portfolio_returns %>%
  mutate(rolling_ff =
            rolling_lm(R_excess,
                       MKT_RF,
                       SMB,
                       HML)) %>%
  slice(-1:-23) %>%
  select(date, rolling_ff)

head(rolling_ff_betas, 3)

# A tibble: 3 x 2
  date        rolling_ff
  <date>      <list>
1 2014-12-31  <S3: lm>
2 2015-01-31  <S3: lm>
3 2015-02-28  <S3: lm>
```

We now have a new data frame called `rolling_ff_betas`, in which the column `rolling_ff` holds an S3 object of our model results. We can `tidy()` that column with `map(rolling_ff, tidy)` and then `unnest()` the results, very similar to our CAPM work except we have more than one independent variable.

```
rolling_ff_betas <-
  ff_portfolio_returns %>%
  mutate(rolling_ff =
            rolling_lm(R_excess,
```

```
                                MKT_RF,
                                SMB,
                                HML)) %>%
    mutate(tidied = map(rolling_ff,
                        tidy,
                        conf.int = T)) %>%
    unnest(tidied) %>%
    slice(-1:-23) %>%
    select(date, term, estimate, conf.low, conf.high) %>%
    filter(term != "(Intercept)") %>%
    rename(beta = estimate, factor = term) %>%
    group_by(factor)

head(rolling_ff_betas, 3)
```

```
# A tibble: 3 x 5
# Groups:   factor [3]
    date        factor      beta conf.low conf.high
    <date>      <chr>      <dbl>    <dbl>     <dbl>
1 2014-12-31 MKT_RF   0.931     0.784    1.08
2 2014-12-31 SMB     -0.0130   -0.278    0.252
3 2014-12-31 HML     -0.160    -0.459    0.139
```

We now have rolling betas and confidence intervals for each of our 3 factors.

We can apply the same code logic to extracting the rolling R-squared for our model. The only difference is we call `glance()` instead of `tidy()`.

```
rolling_ff_rsquared <-
    ff_portfolio_returns %>%
    mutate(rolling_ff =
             rolling_lm(R_excess,
                        MKT_RF,
                        SMB,
                        HML)) %>%
    slice(-1:-23) %>%
    mutate(glanced = map(rolling_ff,
                         glance)) %>%
    unnest(glanced) %>%
    select(date, r.squared, adj.r.squared, p.value)

head(rolling_ff_rsquared, 3)
```

```
# A tibble: 3 x 4
    date        r.squared adj.r.squared  p.value
```

	<date>	<dbl>	<dbl>	<dbl>
1	2014-12-31	0.898	0.883	4.22e-10
2	2015-01-31	0.914	0.901	8.22e-11
3	2015-02-28	0.919	0.907	4.19e-11

We have extracted rolling betas and rolling model results, now let's visualize.

9.4 Visualizing Rolling Fama-French

We start by charting the rolling factor betas with `ggplot()`. This gives us an intuition about how the explanatory power of each factor has changed over time.

```
rolling_ff_betas %>%
  ggplot(aes(x = date,
             y = beta,
             color = factor)) +
  geom_line() +
  labs(title= "24-Month Rolling FF Factor Betas") +
  theme_minimal() +
  theme(plot.title = element_text(hjust = 0.5),
        axis.text.x = element_text(angle = 90))
```

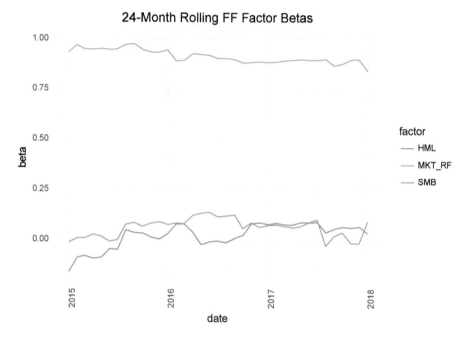

FIGURE 9.2: Rolling Factor Betas

Figure 9.2 reveals some interesting trends. Both SMB and HML have hovered around zero, while the MKT factor has hovered around 1. That's consistent with our plot of betas with confidence intervals.

Now we visualize the rolling R-squared with `highcharter`.

We first convert `rolling_ff_rsquared` to an xts object, using the `tk_xts()` function from `timetk`.

```
rolling_ff_rsquared_xts <-
  rolling_ff_rsquared %>%
  tk_xts(date_var = date, silent = TRUE)
```

Then we pass the `rolling_ff_rsquared_xts` object to a `highchart(type = "stock")` code flow, adding the rolling R-squared time series with `hc_add_series(rolling_ff_rsquared_xts$r.squared...)`.

```
highchart(type = "stock") %>%
  hc_add_series(rolling_ff_rsquared_xts$r.squared,
                color = "cornflowerblue",
                name = "r-squared") %>%
```

```
hc_title(text = "Rolling FF 3-Factor R-Squared") %>%
hc_add_theme(hc_theme_flat()) %>%
hc_navigator(enabled = FALSE) %>%
hc_scrollbar(enabled = FALSE) %>%
hc_exporting(enabled = TRUE)
```

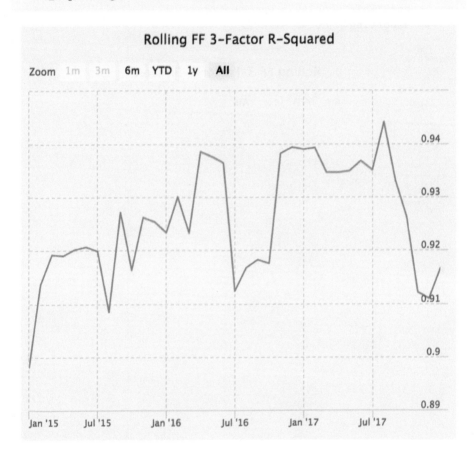

FIGURE 9.3: Rolling FF R-squared highcharter

Figure 9.3 looks choppy but the R-squared never really left the range between
.9 and .95. We can tweak the minimum and maximum y-axis values for some
perspective.

```
highchart(type = "stock") %>%
  hc_add_series(rolling_ff_rsquared_xts$r.squared,
                color = "cornflowerblue",
```

```
                    name = "r-squared") %>%
    hc_title(text = "Rolling FF 3-Factor R-Squared") %>%
    hc_yAxis( max = 2, min = 0) %>%
    hc_add_theme(hc_theme_flat()) %>%
    hc_navigator(enabled = FALSE) %>%
    hc_scrollbar(enabled = FALSE) %>%
    hc_exporting(enabled = TRUE)
```

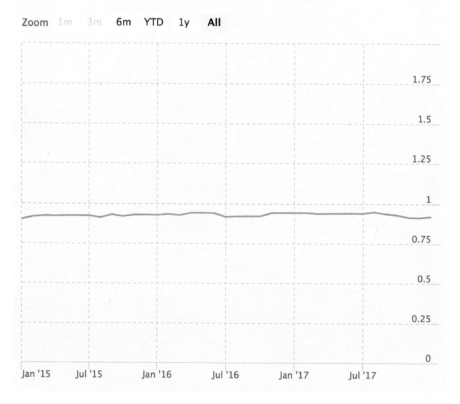

FIGURE 9.4: Rolling R-squared with Y-axis Min Max

As Figure 9.4 shows, when the y-axis is zoomed out a bit, our R-squared looks consistently near 1 for the life of the portfolio.

We have done quite a bit of work to import the FF factors, wrangle them into shape, regress our portfolio returns on them and then pull out the model results for visualization. When we build our Shiny app, we will see how this can be extended to let a user run the analysis on a custom portfolio. We should also think about how this work can be extended to other FF factors or, indeed, any collection of factors.

9.5 Shiny App Fama-French

Now let's port our FF work over to a Shiny dashboard, where a user can build a portfolio to be regressed on the FF factors.

Have a look at the app in Figure 9.5.

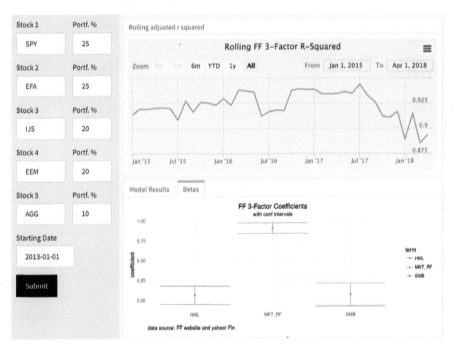

FIGURE 9.5: www.reproduciblefinance.com/shiny/fama-french-three-factor/

Let's start in the **setup** code chunk, which we have previously been using only to load packages. Since we are not allowing the end user any choice about the FF factors, we can import the FF 3-factor data in the **setup**, as soon as the app is accessed.

```r
# This is the setup code chunk for the app.

# Load packages as we do for all Shiny apps
library(tidyquant)
library(tidyverse)
library(timetk)
library(broom)
library(tibbletime)
library(scales)
library(readr)
library(highcharter)

temp <- tempfile()

# Split the url into pieces
base <-
"http://mba.tuck.dartmouth.edu/pages/faculty/ken.french/ftp/"
factor <-
  "Global_3_Factors"
format<-
  "_CSV.zip"

# Paste the pieces together to form the full url
full_url <-
  paste(base,
        factor,
        format,
        sep ="")

download.file(
  # location of file to be downloaded
  full_url,
  # where we want R to store that file
  temp,
  )

Global_3_Factors <-
  read_csv(unz(temp, "Global_3_Factors.csv"),
           skip = 6) %>%
  rename(date = X1) %>%
  mutate_at(vars(-date), as.numeric) %>%
  mutate(date = ymd(parse_date_time(date, "%Y%m")))
```

We calculate portfolio monthly returns and make the change to using the first

day of the month. We have more freedom with Shiny since we are starting
over with creating our returns object.

```
portfolio_returns_tq_rebalanced_monthly_first_day <-
  eventReactive(input$go, {

  prices <- prices()
  w <- c(input$w1/100, input$w2/100,
         input$w3/100, input$w4/100, input$w5/100)

  portfolio_returns_tq_rebalanced_monthly_first_day <-
      prices %>%
      to.monthly(indexAt = "firstof",
                 OHLC = FALSE) %>%
      tk_tbl(preserve_index = TRUE,
             rename_index = "date") %>%
      gather(asset, returns, -date) %>%
      group_by(asset) %>%
      mutate(returns =
               (log(returns) - log(lag(returns)))) %>%
      na.omit() %>%
      tq_portfolio(assets_col = asset,
               returns_col = returns,
               weights     = w,
               col_rename  = "returns",
               rebalance_on = "months")

})
```

Next, we regress those returns on the FF factors and use `tidy(model,
conf.int = T, conf.level = .95)` to get confidence intervals for our betas.
We save the results in `ff_dplyr_byhand()`.

```
ff_dplyr_byhand <- eventReactive(input$go, {

  portfolio_returns_tq_rebalanced_monthly_first_day <-
    portfolio_returns_tq_rebalanced_monthly_first_day()

  ff_dplyr_byhand <-
  portfolio_returns_tq_rebalanced_monthly_first_day %>%
  left_join(Global_3_Factors) %>%
  mutate(MKT = MKT/100,
         SMB = SMB/100,
         HML = HML/100,
```

```
        RF = RF/100,
        Returns_excess = returns - RF ) %>%
   na.omit() %>%
 do(model =
     lm(Returns_excess ~ MKT + SMB + HML, data = .)) %>%
 tidy(model, conf.int = T, conf.level = .95) %>%
 mutate_if(is.numeric, funs(round(., 3))) %>%
 filter(term != "(Intercept)")
})
```

Our final calculation is the rolling r-squared with `rollify()`, which we will
call `rolling_ff_glanced()`.

```
rolling_ff_glanced <- eventReactive(input$go, {

rolling_lm <-
  rollify(.f = function(R_excess, MKT_RF, SMB, HML) {
    lm(R_excess ~ MKT_RF + SMB + HML)
  }, window = input$window, unlist = FALSE)


portfolio_returns_tq_rebalanced_monthly_first_day <-
  portfolio_returns_tq_rebalanced_monthly_first_day()

rolling_ff_glanced <-
  portfolio_returns_tq_rebalanced_monthly_first_day %>%
  left_join(Global_3_Factors, by = "date") %>%
  mutate(MKT_RF = `Mkt-RF`/100,
         SMB = SMB/100,
         HML = HML/100,
         RF = RF/100,
         R_excess = round(returns - RF, 4)) %>%
  mutate(rolling_ff =
           rolling_lm(R_excess, MKT_RF, SMB, HML)) %>%
  slice(-1:-(input$window -1)) %>%
  mutate(glanced = map(rolling_ff,
                       glance)) %>%
  unnest(glanced) %>%
  select(date, r.squared, adj.r.squared, p.value) %>%
  tk_xts(date_var = date, silent = TRUE)

})
```

We pass the rolling xts object to `renderHighchart()`.

```
renderHighchart({

 rolling_ff_glanced <- rolling_ff_glanced()

  highchart(type = "stock") %>%
  hc_add_series(rolling_ff_glanced$adj.r.squared,
              color = "cornflowerblue",
              name = "r-squared") %>%
  hc_title(text = "Rolling FF 3-Factor R-Squared") %>%
  hc_add_theme(hc_theme_flat()) %>%
  hc_navigator(enabled = FALSE) %>%
  hc_scrollbar(enabled = FALSE) %>%
  hc_exporting(enabled = TRUE)
})
```

We pass the model results stored in `ff_dplyr_byhand()` to `renderTable()`.

```
renderTable({
  ff_dplyr_byhand()
})
```

And chart the factor betas, with confidence intervals, by passing `ff_dplyr_byhand()` to `renderPlot()`.

```
renderPlot({
  ff_dplyr_byhand() %>%
  ggplot(aes(x = term,
             y = estimate,
             shape = term,
             color = term)) +
  geom_point() +
  geom_errorbar(aes(ymin = conf.low, ymax = conf.high)) +
  labs(title = "FF 3-Factor Coefficients",
       subtitle = "with conf intervals",
       x = "",
       y = "coefficient",
       caption = "data source: FF website and Yahoo! Fin")
```

That completes our Shiny app for Fama-French, which we can use as a template for multiple linear regression with Shiny and `highcharter`. In general this demonstrates a powerful use of Shiny: we can nest a model or even a rolling model under the hood and then display the results. Once an end user chooses the inputs and clicks submit, we have absolute freedom to utilize R functions in the background and report back to the user with a nice visualization.

Concluding Portfolio Theory

We have reduced a vast and fascinating field to three chapters on Sharpe Ratio, CAPM and Fama-French, explored ratio analysis and linear models, and seen how to run those on a rolling basis. The Fama-French chapter was heavy on data wrangling from a new source because we will frequently be confronted with that situation. If we hypothesize a relationship between different sets of data, we will have to figure out how to make them compatible enough for comparison. Once we have a nice template for doing so, we can quickly explore new data or factor sets for interesting relationships with our portfolio.

If you are starting a new R session and wish to run our code for portfolio theory, first get the portfolio returns objects:

www.reproduciblefinance.com/code/get-returns/

And then see these pages for the risk code flows:

www.reproduciblefinance.com/code/sharpe-ratio

www.reproduciblefinance.com/code/capm-beta/

www.reproduciblefinance.com/code/fama-french-three-factor-model/

Practice and Applications

This section consists of two chapters and two challenges: calculating asset contribution to portfolio standard deviation and running Monte Carlo simulations of returns. Both will involve writing our own functions and deploying them via Shiny. For the first chapter, we do matrix algebra, then use that algebra to create an analytical function, then use that function to create a rolling analytical function, then apply it and fit the results into `highcharter`. In the next chapter, we choose a simulation method, write several functions, run many simulations with `purrr`, then fit the results into `highcharter`.

Those two templates of creating functions and deploying via Shiny, with reproducible and readable code, open up a vast world to us. As long as we can break a task down to matrix algebra or discrete pieces of math, we can wrap those up into our own functions.

By way of a roadmap, we will accomplish the following:

1) write a function to analyze component contribution to standard deviation
2) write a function to analyze *rolling* component contribution to standard deviation
3) visualize component contribution to rolling standard deviation
4) write several functions for simulating future portfolio returns
5) run several simulations with `purrr`
6) visualize the results of those simulations with `highcharter`

From a data science toolkit perspective, the goal is to develop flexible flows and reproducible code for writing our own functions and then deploying them via Shiny.

We will be working with the portfolio returns objects that were created in the *Returns* section. If you are starting a new R session and want to run the code to build those objects, navigate here:

www.reproduciblefinance.com/code/get-returns/

10

Component Contribution to Standard Deviation

Our goal in this chapter is to investigate how each of our 5 assets contributes to portfolio standard deviation. Why might we want to do that?

From a substantive perspective, we may want to ensure that our risk has not got too concentrated in one asset. Not only might this lead to a less diversified portfolio than we intended, but it also might indicate that our initial assumptions about a particular asset were wrong, or at least, have become less right over time.

From an R toolkit perspective, we want to build our own custom function for the job, nest it within another custom function to make it rolling, apply the combination with `purrr` and then use both functions in a Shiny app.

Let's get started.

10.1   Component Contribution Step-by-Step

The percentage contribution to portfolio standard deviation of an asset is defined as:

(marginal contribution of the asset * weight of the asset) / portfolio standard deviation

We start by building the covariance matrix and calculating portfolio standard deviation.

```
covariance_matrix <-
  cov(asset_returns_xts)

sd_portfolio <-
  sqrt(t(w) %*% covariance_matrix %*% w)
```

We find the marginal contribution of each asset by taking the cross product of the weights vector and the covariance matrix, divided by the portfolio standard deviation.

```
marginal_contribution <-
  w %*% covariance_matrix / sd_portfolio[1, 1]
```

Now we multiply the marginal contribution of each asset by our original weights vector w, which we defined way back in the *Returns* section, to get total contribution.

```
component_contribution <-
  marginal_contribution * w
```

We can sum the asset contributions and check that the result equals total portfolio standard deviation.

```
components_summed <- rowSums(component_contribution)

components_summed
```

```
[1] 0.02661
```

```
sd_portfolio[1,1]
```

```
[1] 0.02661
```

The summed components are equal to the total as we hoped.

To get to percentage contribution of each asset, we divide each asset's contribution by total portfolio standard deviation.

```
component_percentages <-
  component_contribution / sd_portfolio[1, 1]

round(component_percentages, 3)
```

```
       SPY    EFA    IJS  EEM    AGG
[1,] 0.233 0.276 0.227 0.26 0.003
```

That is the step-by-step code flow, now let's get functional.

10.2 Component Contribution with a Custom Function

Notice in the above code that we supplied only two pieces of data: asset_returns_xts and w. Thus, we can write a function that takes two arguments, an individual asset returns object and a weights vector, and then calculates the contribution to standard deviation of each of the assets. We title our function component_contr_matrix_fun() and port our previous step-by-step code.

```r
component_contr_matrix_fun <- function(returns, w){
# create covariance matrix
covariance_matrix <-
  cov(returns)
# calculate portfolio standard deviation
sd_portfolio <-
  sqrt(t(w) %*% covariance_matrix %*% w)
# calculate marginal contribution of each asset
marginal_contribution <-
  w %*% covariance_matrix / sd_portfolio[1, 1]
# multiply marginal by weights vecotr
component_contribution <-
  marginal_contribution * w
# divide by total standard deviation to get percentages
component_percentages <-
  component_contribution / sd_portfolio[1, 1]

component_percentages %>%
  as_tibble() %>%
  gather(asset, contribution)
}
```

Note the final two lines in that function. as_tibble() converts the result to a tibble and gather(asset, contribution) converts to a tidy format.

Let's test the function and confirm it returns the same results as our step-by-step code flow if we pass in the same data as before: asset_returns_xts and w.

```r
test_the_function_xts <-
  component_contr_matrix_fun(asset_returns_xts, w)

test_the_function_xts
```

```
# A tibble: 5 x 2
  asset contribution
  <chr>        <dbl>
1 SPY          0.233
2 EFA          0.276
3 IJS          0.227
4 EEM          0.260
5 AGG          0.00318
```

The substantive results are the same and they are in a tidy `tibble` now.

Let's test whether we can pass in a `tibble` of returns. Note that we need to remove the date column first with `select(-date)`, else the function will think the date column is another column of asset returns.

```
percentages_tibble <-
  asset_returns_dplyr_byhand %>%
  select(-date) %>%
  component_contr_matrix_fun(., w)

percentages_tibble
```

```
# A tibble: 5 x 2
  asset contribution
  <chr>        <dbl>
1 SPY          0.233
2 EFA          0.276
3 IJS          0.227
4 EEM          0.260
5 AGG          0.00318
```

We have consistent results from our step-by-step code, and our custom function when passing in `xts` and `tibble` data.

That custom function accomplishes what we want, but we are not the first ones to imagine the usefulness of component contribution to standard deviation. There is a built-in function to accomplish this with `StdDev(asset_returns_xts, weights = w, portfolio_method = "component")` from the `PerformanceAnalytics` package.

However, writing our own function allows us flexibility: we were able to pass in a `tibble` and specify the return format as a tidy `tibble`, we could alter the matrix algebra and calculate an entirely different statistic, we could alter the output to include the weights as a column. Our creativity is the only limitation with custom functions, especially once we have a template for writing and creating them.

Next we will see how returning a `tibble` facilitates visualization.

10.3 Visualizing Component Contribution

The output of our custom function, `percentages_tibble`, is already tidy. We can pipe to `ggplot()` and use `geom_bar()` to chart risk contribution on the y-axis.

```
percentages_tibble %>%
  ggplot(aes(x = asset, y = contribution)) +
  geom_col(fill = 'cornflowerblue',
           colour = 'pink',
           width = .6) +
  scale_y_continuous(labels = percent,
                     breaks = pretty_breaks(n = 20)) +
  ggtitle("Percent Contribution to Standard Deviation") +
  theme(plot.title = element_text(hjust = 0.5)) +
  xlab("Asset") +
  ylab("Percent Contribution to Risk")
```

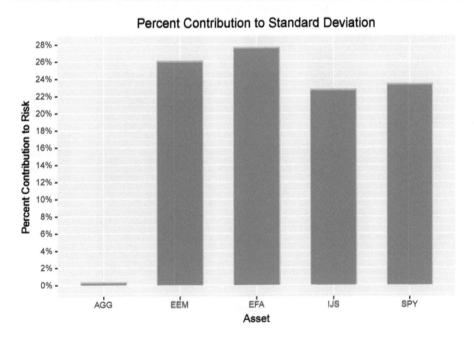

FIGURE 10.1: Contribution to Standard Deviation

Figure 10.1 shows the individual contributions but it would be useful to see a chart that compares asset weight to risk contribution.

For that, we gather our `tibble` to long format with `gather(type, percent, -asset)`, then call `ggplot(aes(x = asset, y = percent, fill = type))`. Position the columns so that they are not right on top of each other with `geom_col(position='dodge')`.

```
percentages_tibble %>%
mutate(weights = w) %>%
gather(type, percent, -asset) %>%
group_by(type) %>%
ggplot(aes(x = asset,
          y = percent,
          fill = type)) +
geom_col(position='dodge') +
scale_y_continuous(labels = percent) +
ggtitle("Percent Contribution to Volatility") +
theme(plot.title = element_text(hjust = 0.5))
```

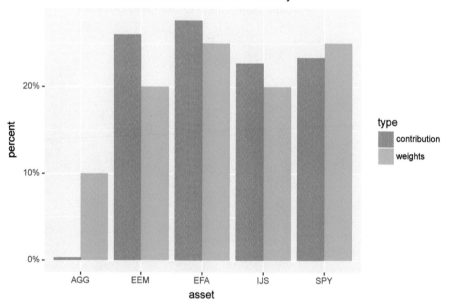

FIGURE 10.2: Weight versus Contribution

Figure 10.2 makes it clear that AGG, the bond fund, has done a good job as a volatility dampener. It has a 10% allocation but contributes almost zero to volatility.

We have written a custom function and used the results in a nice visualization. Think about how we could add the `ggplot()` flow to the custom function, so that the result is a data visualization.

10.4 Rolling Component Contribution

Now we turn to our most involved task yet, which is to calculate the rolling volatility contribution of each asset. The previous section told us the total contribution of each asset over the life of the portfolio, but it did not help us understand risk components over time.

To gain that understanding, we need to create a function that calculates the rolling contribution to standard deviation after being supplied with 4 arguments:

```
1) asset returns
2) portfolio weights
3) starting date
4) rolling window
```

Before we start coding, here is the logic that we wish to implement (feel free to eviscerate this and replace it with something better).

1. Assign a start date and rolling window.
2. Assign an end date based on the start date and window. If we set window = 24, we will be calculating contributions for the 24-month period between the start date and the end date.
3. Use `filter()` to subset the original data. I label the subsetted data frame as `returns_to_use`.
4. Assign a weights object called `w` based on the weights argument.
5. Pass `returns_to_use` and `w` to `component_contr_matrix_fun()`, the function we created in section 10.2. We are nesting one custom function inside another.
6. After running `component_contr_matrix_fun()`, we have an object called `component_percentages`. What is this? It is the risk contribution for each asset during the first 24-month window.
7. Add a date to `component_percentages` with `mutate(date = ymd(end_date))`. We have the component contributions as of the end date.
8. We now have the risk contributions for the 24-month period that started on the first date, or January 2013, and ended on the end

date, January 2015, because we default to `start = 1`. If we wanted to get the risk contribution for a 24-month period that started on the second date or February 2013, we would set `start = 2`, etc.

9. If we want risk contribution for all the 24-month periods, we need to apply this function starting at January, then February, then March, then April, all the way to the start date that is 24 months before the end of our data.

Now, we turn those enumerated steps into a function.

```
interval_sd_by_hand <-
  function(returns_df,
           start = 1,
           window = 24,
           weights){

  # First create start date.
  start_date <-
    returns_df$date[start]

  # Next create an end date that depends
  # on start date and window.
  end_date <-
    returns_df$date[c(start + window)]

  # Filter on start and end date.
  returns_to_use <-
    returns_df %>%
    filter(date >= start_date & date < end_date) %>%
    select(-date)

  # Portfolio weights
  w <- weights

  # Call our original custom function
  # We are nesting one function inside another
  component_percentages <-
   component_contr_matrix_fun(returns_to_use, w)

  # Add back the end date as date column
  results_with_date <-
    component_percentages %>%
    mutate(date = ymd(end_date)) %>%
    select(date, everything()) %>%
```

```
    spread(asset, contribution) %>%
    # Round the results for better presentation
    mutate_if(is.numeric, function(x) x * 100)
}
```

We can test our work and apply that function to a returns `tibble`.

```
test_interval_function_1 <-
  interval_sd_by_hand(asset_returns_dplyr_byhand,
                      start = 1,
                      window = 24,
                      weights = w)


test_interval_function_1
```

```
# A tibble: 1 x 6
  date          AGG   EEM   EFA   IJS   SPY
  <date>      <dbl> <dbl> <dbl> <dbl> <dbl>
1 2015-01-31   1.16  25.6  29.2  23.0  21.1
```

The function returns the contribution to risk for each asset from January 2013 through January, 2015. The reason the date column is 2015-01-31 is that, as of that date, our results are the contribution over the previous 24 months of each asset.

What if we started at `start = 2`?

```
test_interval_function_2 <-
  interval_sd_by_hand(asset_returns_dplyr_byhand,
                      start = 2,
                      window = 24,
                      weights = w) %>%
      mutate_if(is.numeric, funs(round(., 3)))


test_interval_function_2
```

```
# A tibble: 1 x 6
  date          AGG   EEM   EFA   IJS   SPY
  <date>      <dbl> <dbl> <dbl> <dbl> <dbl>
1 2015-02-28   1.00  25.6  28.0  23.9  21.5
```

We have moved up by one month to February, 2015.

We could keep doing this manually, moving up one date with each function call, and then paste together all of the results.

Instead, we will use the `purrr` package to apply `interval_sd_by_hand()` and demonstrate the power of `purrr` plus writing custom functions.

`purrr` contains a family of `map` functions, which apply another function iteratively, similar to a loop. Here, we call `map_df()` to apply our rolling function iteratively, looping over the date index of our returns object. The appending of `_df` tells `map()` to store our results in `tibble` rows.

`map_df()` takes 5 arguments: the function to be applied, which is `interval_sd_by_hand()`, and the 4 arguments to that function, which are returns, weights, a rolling window and the starting date index.

```
window <- 24

portfolio_vol_components_tidy_by_hand <-
  # First argument:
  # tell map_df to start at date index 1
  # This is the start argument to interval_sd_by_hand()
  # and it is what map() will loop over until we tell
  # it to stop at the date that is 24 months before the
  # last date.
  map_df(1:(nrow(asset_returns_dplyr_byhand) - window),
         # Second argument:
         # tell it to apply our rolling function
         interval_sd_by_hand,
         # Third argument:
         # tell it to operate on our returns
         returns_df = asset_returns_dplyr_byhand,
         # Fourth argument:
         # supply the weights
         weights = w,
         # Fifth argument:
         # supply the rolling window
         window = window)

tail(portfolio_vol_components_tidy_by_hand)
```

```
# A tibble: 6 x 6
  date           AGG   EEM   EFA   IJS   SPY
  <date>        <dbl> <dbl> <dbl> <dbl> <dbl>
1 2017-07-31  0.133   26.8  26.4  22.4  24.2
2 2017-08-31  0.182   26.6  27.0  21.8  24.5
3 2017-09-30 -0.0396  25.6  26.4  23.9  24.2
4 2017-10-31  0.0321  25.3  25.7  24.8  24.2
5 2017-11-30  0.0853  26.7  25.4  25.7  22.1
6 2017-12-31  0.0138  26.1  25.2  26.4  22.3
```

The result is the rolling component contribution to standard deviation of our 5 assets. That last date is December 31, 2017. The results for that date are the contribution to standard deviation for each asset over the preceding 24 months.

We did quite a bit of work to get that data: we ran a calculation with matrix algebra step-by-step, then wrapped it into a function called `component_contr_matrix_fun()`, then created another function called `my_interval_sd()` to make it rolling, then applied it with `map_df()`. Now we know why R coders spend so much time at white boards!

Now let's complete our task by visualizing these results.

·10.5   Visualizing Rolling Component Contribution

We begin our aesthetic journey with `ggplot()`. Since our results are in wide format, we first conver them to long format with `gather(asset, contribution, -date)` and then chart by asset.

```
portfolio_vol_components_tidy_by_hand %>%
  gather(asset, contribution, -date) %>%
  group_by(asset) %>%
  ggplot(aes(x = date)) +
  geom_line(aes(y = contribution,
                color = asset)) +
  scale_x_date(breaks =
                 pretty_breaks(n = 8)) +
  scale_y_continuous(labels =
                 function(x) paste0(x, "%"))
```

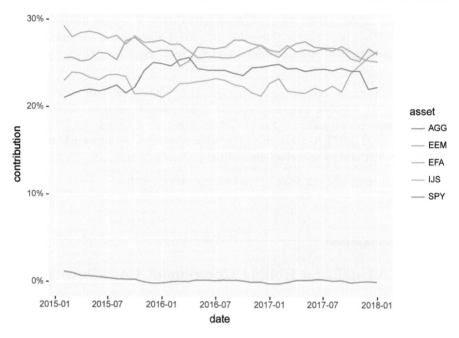

FIGURE 10.3: Component Contribution ggplot

Figure 10.3 clearly conveys how the 5 assets have contributed over time. In particular, when one asset's contribution surpasses another asset's contribution, we can see where the lines cross.

That said, percentages that total to 100 are very often visualized with a stacked chart in the financial world. We build such a chart with `geom_area(aes(colour = asset, fill= asset), position = 'stack')`, which tells `ggplot()` to construct an area chart where each geom is stacked.

```
portfolio_vol_components_tidy_by_hand %>%
    gather(asset, contribution, -date) %>%
    group_by(asset) %>%
ggplot(aes(x = date,
           y = contribution)) +
geom_area(aes(colour = asset,
              fill= asset),
          position = 'stack') +
  scale_x_date(breaks =
                   pretty_breaks(n = 8)) +
  scale_y_continuous(labels =
                   function(x) paste0(x, "%"))
```

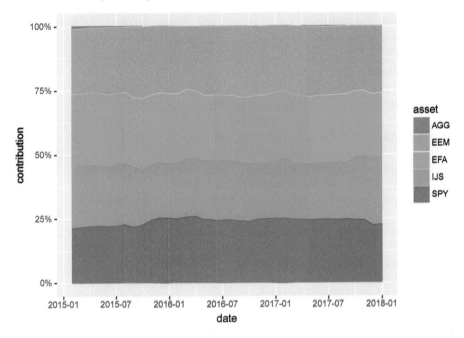

FIGURE 10.4: Stacked Component Contribution ggplot

In Figure 10.4, `ggplot()` has stacked these in alphabetical order and it is hard to see AGG's contribution at all. Stacked charts are common in finance when time series sum to 100 but I find that they convey less information than our original line charts.

Let's head to `highcharter` and recreate those rolling contribution charts.

First, we need to convert the `tibble` to an `xts` with `tk_xts(date_var = date)`.

```
portfolio_vol_components_tidy_xts <-
  portfolio_vol_components_tidy_by_hand %>%
  tk_xts(date_var = date,
        silent = TRUE)
```

Now we add each time series to the `highchart()` flow.

Note that we tweak the minimum and maximum y-axis values by adding `hc_yAxis(...max = max(portfolio_vol_components_tidy_xts) + 5, min = min(portfolio_vol_components_tidy_xts) - 5`. That line of code will set the maximum y-axis value to the highest value in our data

plus 5 and the minimum y-axis value to the lowest value in our data minus 5.
This is purely aesthetic.

```r
highchart(type = "stock") %>%
hc_title(text = "Volatility Contribution") %>%
hc_add_series(portfolio_vol_components_tidy_xts[, 1],
              name = symbols[1]) %>%
hc_add_series(portfolio_vol_components_tidy_xts[, 2],
              name = symbols[2]) %>%
hc_add_series(portfolio_vol_components_tidy_xts[, 3],
              name = symbols[3]) %>%
hc_add_series(portfolio_vol_components_tidy_xts[, 4],
              name = symbols[4]) %>%
hc_add_series(portfolio_vol_components_tidy_xts[, 5],
              name = symbols[5]) %>%
hc_yAxis(labels = list(format = "{value}%"),
    max = max(portfolio_vol_components_tidy_xts) + 5,
    min = min(portfolio_vol_components_tidy_xts) - 5,
    opposite = FALSE) %>%
hc_navigator(enabled = FALSE) %>%
hc_scrollbar(enabled = FALSE) %>%
hc_add_theme(hc_theme_flat()) %>%
hc_exporting(enabled = TRUE) %>%
hc_legend(enabled = TRUE)
```

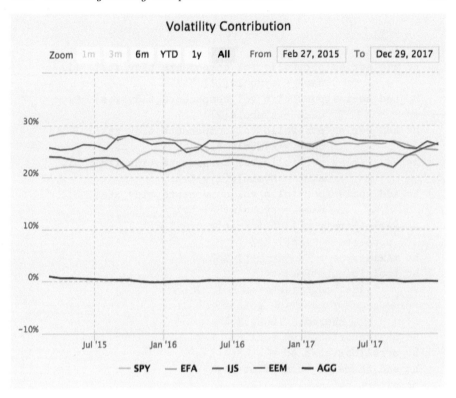

FIGURE 10.5: Rolling Contribution highcharter

Figure 10.5 looks good and we can see clearly that the IJS has been contributing more to volatility recently.

To build a stacked chart using `highcharter`, we first change to an area chart with `hc_chart(type = "area")` and then set `hc_plotOptions(area = list(stacking = "percent"...)`.

```
highchart() %>%
  hc_chart(type = "area") %>%
  hc_title(text = "Volatility Contribution") %>%
  hc_plotOptions(area = list(
    stacking = "percent",
    lineColor = "#ffffff",
    lineWidth = 1,
    marker = list(
      lineWidth = 1,
      lineColor = "#ffffff"
```

```
        ))
      ) %>%
  hc_add_series(portfolio_vol_components_tidy_xts[, 1],
                name = symbols[1]) %>%
  hc_add_series(portfolio_vol_components_tidy_xts[, 2],
                name = symbols[2]) %>%
  hc_add_series(portfolio_vol_components_tidy_xts[, 3],
                name = symbols[3]) %>%
  hc_add_series(portfolio_vol_components_tidy_xts[, 4],
                name = symbols[4]) %>%
  hc_add_series(portfolio_vol_components_tidy_xts[, 5],
                name = symbols[5]) %>%
  hc_yAxis(labels = list(format = "{value}%"),
      opposite = FALSE) %>%
  hc_xAxis(type = "datetime") %>%
  hc_tooltip(pointFormat =
"<span style=\"color:{series.color}\">
{series.name}</span>:<b>{point.percentage:.1f}%</b><br/>",
              shared = TRUE) %>%
  hc_navigator(enabled = FALSE) %>%
  hc_scrollbar(enabled = FALSE) %>%
  hc_add_theme(hc_theme_flat()) %>%
  hc_exporting(enabled = TRUE) %>%
  hc_legend(enabled = TRUE)
```

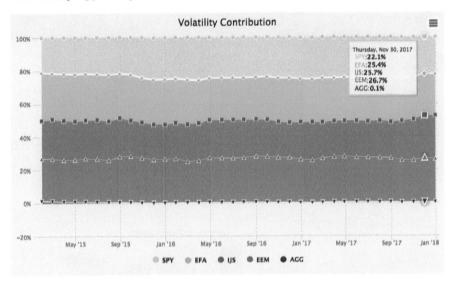

FIGURE 10.6: Stacked Component Contribution highcharter

Figure 10.6 is more compelling on the internet where you can hover on different points and see the tooltip change over time. You can view it here:

www.reproduciblefinance.com/code/rolling-component-contribution-highcharter/

10.6 Shiny App Component Contribution

Finally, we deploy a Shiny app that allows an end user to build a custom portfolio, select a rolling window, and chart the rolling contributions to risk of each asset.

Have a look at the app in Figure 10.7:

FIGURE 10.7: www.reproduciblefinance.com/shiny/volatility-contribution/

Note that we are including 3 data visualizations from our previous work.

1) a `highcharter` of rolling asset contributions
2) a `highcharter` stacked area chart
3) a `ggplot()` bar chart of asset weights and contributions

This app uses our standard input sidebar that lets the user choose stocks, weights and a start date, plus a rolling window.

Very importantly, we then use our custom functions `component_contr_matrix_fun` and `interval_sd_by_hand()`. To do that in Shiny, we define the function the same as we would in a normal R script. Any function that we can create in R can be used in a Shiny app and we define it in the same way. That is the power of Shiny for R coders. At the beginning of each project, we know that whatever creative code, models, and visualizations we build, those can be turned into an interactive web application.

Here's the code for our custom functions. It's identical to how we coded them above. What changes is that they will be passed reactive values downstream.

```
interval_sd_by_hand <-
  function(returns_df,
           start = 1,
           window = 24,
           weights){

  # First create start date.
  start_date <-
    returns_df$date[start]

  # Next an end date that depends on start date and window.
  end_date <-
    returns_df$date[c(start + window)]

  # Filter on start and end date.
  returns_to_use <-
    returns_df %>%
    filter(date >= start_date & date < end_date) %>%
    select(-date)

  # Portfolio weights.
  w <- weights

  # Call our original custom function.
  # We are nesting one function inside another.
  component_percentages <-
   component_contr_matrix_fun(returns_to_use, w)

  # Add back the end date as date column
  results_with_date <-
    component_percentages %>%
    mutate(date = ymd(end_date)) %>%
    select(date, everything()) %>%
    spread(asset, contribution) %>%
    mutate_if(is.numeric, function(x) x * 100)
}
```

Then we calculate asset returns using an eventReactive().

```
asset_returns_dplyr_byhand <- eventReactive(input$go, {
```

```
  symbols <- c(input$stock1, input$stock2,
               input$stock3, input$stock4, input$stock5)

  prices <-
    getSymbols(symbols, src = 'yahoo', from = input$date,
               auto.assign = TRUE, warnings = FALSE) %>%
    map(~Ad(get(.))) %>%
    reduce(merge) %>%
    `colnames<-`(symbols)

  asset_returns_dplyr_byhand <-
    prices %>%
    to.monthly(indexAt = "last", OHLC = FALSE) %>%
    tk_tbl(preserve_index = TRUE, rename_index = "date") %>%
    gather(asset, returns, -date) %>%
    group_by(asset) %>%
    mutate(returns = (log(returns) - log(lag(returns)))) %>%
    spread(asset, returns) %>%
    select(date, symbols) %>%
    slice(-1)
})
```

Now we can call our custom function to calculate overall contribution to risk.

```
percentages_tibble_pre_built <- eventReactive(input$go, {

  asset_returns_xts <-
    asset_returns_dplyr_byhand() %>%
    tk_xts(date_col = date)

  w <- c(input$w1/100, input$w2/100,
         input$w3/100, input$w4/100, input$w5/100)

  port_vol_contr_total_builtin <-
    StdDev(asset_returns_xts,
           weights = w,
           portfolio_method = "component")

  symbols <- c(input$stock1, input$stock2,
               input$stock3, input$stock4, input$stock5)

  percentages_tibble_pre_built <-
    port_vol_contr_total_builtin$pct_contrib_StdDev %>%
```

```
   tk_tbl(preserve_index = FALSE) %>%
   mutate(asset = symbols) %>%
   rename('risk contribution' = data) %>%
   mutate(`risk contribution` =
          round(`risk contribution`, 4) * 100,
        weights = w * 100) %>%
   select(asset, everything())

})
```

And finally our major calculation for this Shiny app: calculate rolling contribution to standard deviation using our hand-built function, and apply it with map_df().

```
portfolio_vol_components_xts <- eventReactive(input$go, {

  asset_returns_dplyr_byhand <- asset_returns_dplyr_byhand()

  w <- c(input$w1/100, input$w2/100,
         input$w3/100, input$w4/100,
         input$w5/100)

portfolio_vol_components_tidy_by_hand <-
  map_df(1:(nrow(asset_returns_dplyr_byhand) - input$window),
         interval_sd_by_hand,
         returns_df = asset_returns_dplyr_byhand,
         weights = w,
         window = input$window) %>%
  tk_xts(date_col = date)
})
```

From here we visualize first with highcharter.

```
renderHighchart({
  portfolio_vol_components <-
    portfolio_vol_components_xts()

  highchart(type = "stock") %>%
    hc_title(text = "Volatility Contribution") %>%
    hc_add_series(portfolio_vol_components[, 1],
            name = names(portfolio_vol_components[, 1])) %>%
    hc_add_series(portfolio_vol_components[, 2],
            name = names(portfolio_vol_components[, 2])) %>%
```

```
    hc_add_series(portfolio_vol_components[, 3],
            name = names(portfolio_vol_components[, 3])) %>%
    hc_add_series(portfolio_vol_components[, 4],
            name = names(portfolio_vol_components[, 4])) %>%
    hc_add_series(portfolio_vol_components[, 5],
            name = names(portfolio_vol_components[, 5])) %>%
    hc_add_theme(hc_theme_flat()) %>%
    hc_yAxis(
      labels = list(format = "{value}%"),
            opposite = FALSE,
            min = min(portfolio_vol_components) -5,
            max = max(portfolio_vol_components) + 5) %>%
    hc_navigator(enabled = FALSE) %>%
    hc_scrollbar(enabled = FALSE)
})
```

We also give the user the option of clicking on a stacked area chart.

```
renderHighchart({
  portfolio_vol_components <-
    portfolio_vol_components_xts()

 highchart() %>%
  hc_chart(type = "area") %>%
  hc_title(text = "Stacked Volatility Contribution") %>%
  hc_plotOptions(area = list(
    stacking = "percent",
    lineColor = "#ffffff",
    lineWidth = 1,
    marker = list(
      lineWidth = 1,
      lineColor = "#ffffff"
      ))
    ) %>%
    hc_add_series(portfolio_vol_components[, 1],
            name = names(portfolio_vol_components[, 1])) %>%
    hc_add_series(portfolio_vol_components[, 2],
            name = names(portfolio_vol_components[, 2])) %>%
    hc_add_series(portfolio_vol_components[, 3],
            name = names(portfolio_vol_components[, 3])) %>%
    hc_add_series(portfolio_vol_components[, 4],
            name = names(portfolio_vol_components[, 4])) %>%
    hc_add_series(portfolio_vol_components[, 5],
```

```
              name = names(portfolio_vol_components[, 5])) %>%
      hc_yAxis(labels = list(format = "{value}%"),
        opposite = FALSE) %>%
    hc_xAxis(type = "datetime") %>%
    hc_tooltip(pointFormat =
  "<span style=\"color:{series.color}\">
  {series.name}</span>:<b>{point.percentage:.1f}%</b><br/>",
              shared = TRUE) %>%
    hc_navigator(enabled = FALSE) %>%
    hc_scrollbar(enabled = FALSE) %>%
    hc_add_theme(hc_theme_flat()) %>%
    hc_exporting(enabled = TRUE) %>%
    hc_legend(enabled = TRUE)
})
```

Lastly, we build the bar chart showing weights versus contribution to volatility using `percentages_tibble_pre_built()` and `renderPlot`.

```
renderPlot({
  percentages_tibble_pre_built() %>%
  gather(type, percent, -asset) %>%
  group_by(type) %>%
  mutate(percent = percent/100) %>%
  ggplot(aes(x = asset, y = percent, fill = type)) +
  geom_col(position = 'dodge') +
  scale_y_continuous(labels = percent) +
  ggtitle("Percent Contribution to Volatility") +
  theme(plot.title = element_text(hjust = 0.5))
})
```

That completes our Shiny app for visualizing asset contribution to portfolio standard deviation. Think about how we might build a similar app for a portfolio with 50 or 100 assets. The end user would not want to enter 50 tickers and weights, and even if the end user were willing, the probability of making an error is high. How could we make that app usable? Perhaps allow the user to upload a csv with tickers and weights, in a bring-your-own-data scenario? It's a good challenge for the future.

11

Monte Carlo Simulation

We conclude our project by exploring a Monte Carlo (MC) method for simulating future portfolio returns. MC relies on repeated, random sampling from a distribution and we will create that distribution based on two parameters: mean and standard deviation of returns. This is a simple MC method that we are using as a vehicle to introduce code flows and functions. As with the previous chapter, the goal is to demonstrate a reproducible simulation, from building functions through to Shiny deployment, that can serve as the basis for more complex work.

By way of specific tasks we will do the following:

1) write several simulation functions
2) run several simulations with `purrr`
3) visualize our results with `highcharter` and `ggplot2`

We will be working with the portfolio returns objects that were created in the *Returns* section. If you are starting a new R session and want to run the code to build those objects, navigate here:

www.reproduciblefinance.com/code/get-returns/

11.1   Simulating Growth of a Dollar

To simulate based on mean and standard deviation, we first assign our mean and standard deviation of returns to variables called `mean_port_return` and `stddev_port_return`. Those names might seem wordy but we never want to assign a variable with the name `mean` because `mean` is also the name of a function in R.

```
mean_port_return <-
  mean(portfolio_returns_tq_rebalanced_monthly$returns)
```

```
stddev_port_return <-
  sd(portfolio_returns_tq_rebalanced_monthly$returns)
```

Then we use the `rnorm()` function to create a distribution with mean equal to `mean_port_return` and standard deviation equal to `stddev_port_return`. That is the crucial random returns generating step that underpins this entire chapter. Again, it is a simple MC method.

We also must decide how many observations we will simulate for this distribution, meaning how many monthly returns we will simulate. One hundred twelve months is ten years and that feels like a reasonable amount of time.

```
simulated_monthly_returns <- rnorm(120,
                                   mean_port_return,
                                   stddev_port_return)
```

Have a quick look at the simulated monthly returns.

```
head(simulated_monthly_returns)
```

```
[1] -0.029711  0.002365  0.033967 -0.010560 -0.010957
[6]  0.002426
```

```
tail(simulated_monthly_returns)
```

```
[1]  0.002677 -0.002628  0.006577  0.022055  0.009889
[6] -0.008980
```

Next, we calculate how a dollar would have grown given those random monthly returns. We first add a 1 to each of our monthly returns, because we start with $1.

```
simulated_returns_add_1 <-
  tibble(c(1, 1 + simulated_monthly_returns)) %>%
  `colnames<-`("returns")
```

```
head(simulated_returns_add_1)
```

```
# A tibble: 6 x 1
  returns
    <dbl>
1   1.00
2   0.970
```

3	1.00
4	1.03
5	0.989
6	0.989

That data is now ready to be converted into the cumulative growth of a dollar. We can use either `accumulate()` from `purrr` or `cumprod()`. Let's use both of them with `mutate()` and confirm consistent, reasonable results.

```
simulated_growth <-
simulated_returns_add_1 %>%
    mutate(growth1 = accumulate(returns, function(x, y) x * y),
           growth2 = accumulate(returns, `*`),
           growth3 = cumprod(returns)) %>%
    select(-returns)


tail(simulated_growth)
```

```
# A tibble: 6 x 3
  growth1 growth2 growth3
    <dbl>   <dbl>   <dbl>
1    2.10    2.10    2.10
2    2.10    2.10    2.10
3    2.11    2.11    2.11
4    2.16    2.16    2.16
5    2.18    2.18    2.18
6    2.16    2.16    2.16
```

We just ran 3 simulations of dollar growth over 120 months. We passed in the same monthly returns and that's why we got 3 equivalent results.

Are they reasonable? What compound annual growth rate (CAGR) is implied by this simulation?

```
cagr <-
  ((simulated_growth$growth1[nrow(simulated_growth)]^
      (1/10)) -1)*100
```

This simulation implies an annual compounded growth of 8.0042%. That seems reasonable since our actual returns have all been taken from a raging bull market. Remember, the above code is a simulation based on sampling from a normal distribution. If you re-run this code on your own, you will get a different result.

If we feel good about this first simulation, we can run more to get a sense for how they are distributed. Before we do that, let's create several different functions that could run the same simulation.

11.2   Several Simulation Functions

Let's build 3 simulation functions that incorporate the `accumulate()` and `cumprod()` workflows above. We have confirmed they give consistent results so it's a matter of stylistic preference as to which one is chosen in the end. Perhaps you feel that one is more flexible or extensible or fits better with your team's code flows.

Four arguments are required for each of the simulation functions: N for the number of months to simulate (we chose 120 above), `init_value` for the starting value (we used $1 above) and the mean-standard deviation pair to create a normal distribution. We *choose* N and `init_value`, and derive the mean-standard deviation pair from our portfolio monthly returns.

Here is our first growth simulation function using `accumulate()`.

```
simulation_accum_1 <- function(init_value, N, mean, stdev) {
    tibble(c(init_value, 1 + rnorm(N, mean, stdev))) %>%
    `colnames<-`("returns") %>%
    mutate(growth =
              accumulate(returns,
                         function(x, y) x * y)) %>%
    select(growth)
}
```

Here is a second almost identical simulation function using `accumulate()`. Why am I including both? People have different stylistic and aesthetic preferences, even for code.

```
simulation_accum_2 <- function(init_value, N, mean, stdev) {
   tibble(c(init_value, 1 + rnorm(N, mean, stdev))) %>%
    `colnames<-`("returns") %>%
   mutate(growth = accumulate(returns, `*`)) %>%
   select(growth)
}
```

Finally, here is a simulation function using `cumprod()`.

```
simulation_cumprod <- function(init_value, N, mean, stdev) {
   tibble(c(init_value, 1 + rnorm(N, mean, stdev))) %>%
    `colnames<-`("returns") %>%
   mutate(growth = cumprod(returns)) %>%
```

```
  select(growth)
}
```

Here is a function that uses those three previous functions, for a fast way to re-confirm consistency.

```
simulation_confirm_all <- function(init_value, N, mean, stdev) {
  tibble(c(init_value, 1 + rnorm(N, mean, stdev))) %>%
    `colnames<-`("returns") %>%
    mutate(growth1 = accumulate(returns, function(x, y) x * y),
           growth2 = accumulate(returns, `*`),
           growth3 = cumprod(returns)) %>%
    select(-returns)
}
```

Let's test that `confirm_all()` function with an `init_value` of 1, N of 120, and our parameters.

```
simulation_confirm_all_test <-
  simulation_confirm_all(1, 120,
                         mean_port_return, stddev_port_return)

tail(simulation_confirm_all_test)
```

```
# A tibble: 6 x 3
  growth1 growth2 growth3
    <dbl>   <dbl>   <dbl>
1    4.55    4.55    4.55
2    4.51    4.51    4.51
3    4.56    4.56    4.56
4    4.51    4.51    4.51
5    4.43    4.43    4.43
6    4.41    4.41    4.41
```

Now we are ready to run more than one simulation using the function of our choice.

11.3   Running Multiple Simulations

Let's suppose we wish to run 51 simulations because they are random and we want to get a feel for how the randomness is distributed.

First, we need an empty matrix with 51 columns, an initial value of $1 and intuitive column names.

We will use the `rep()` function to create 51 columns with a 1 as the value and `set_names()` to name each column with the appropriate simulation number.

```
sims <- 51
starts <-
  rep(1, sims) %>%
  set_names(paste("sim", 1:sims, sep = ""))
```

Take a peek at `starts` to see what we just created and how it can house our simulations.

```
head(starts)
```

```
sim1 sim2 sim3 sim4 sim5 sim6
   1    1    1    1    1    1
```

```
tail(starts)
```

```
sim46 sim47 sim48 sim49 sim50 sim51
    1     1     1     1     1     1
```

We now have 51 columns, each of which has a value of 1. This is where we will store the results of the 51 simulations.

Now we want to apply one of our simulation functions (we will go with `simulation_accum_1`) to each of the 51 columns of the `starts` matrix. We will do that using the `map_dfc()` function from the `purrr` package.

`map_dfc()` takes a vector, in this case the columns of `starts`, and applies a function to it. By appending `dfc()` to the `map_` function, we are asking the function to store each of its results as the column of a data frame (recall that `map_df()` does the same thing, but stores results in the rows of a data frame). After running the code below, we will have a data frame with 51 columns, one for each of our simulations.

We still need to choose how many months to simulate (the N argument to our simulation function) and supply the distribution parameters as we did before. We do not supply the `init_value` argument because the `init_value` is 1, that same 1 which is in the 51 columns.

```
monte_carlo_sim_51 <-
  map_dfc(starts, simulation_accum_1,
          N = 120,
          mean = mean_port_return,
          stdev = stddev_port_return)

tail(monte_carlo_sim_51 %>%
       select(growth1, growth2,
              growth49, growth50), 3)
```

```
# A tibble: 3 x 4
  growth1 growth2 growth49 growth50
    <dbl>   <dbl>    <dbl>    <dbl>
1    3.12    2.44     2.51     3.85
2    3.17    2.33     2.52     3.87
3    3.26    2.34     2.59     3.92
```

Have a look at the results. We now have 51 simulations of the growth of a dollar and we simulated that growth over 120 months, but the results are missing a piece that we need for visualization, namely a `month` column.

Let's add that `month` column with `mutate()` and give it the same number of rows as our data frame. These are months out into the future. We will use `mutate(month = seq(1:nrow(.)))` and then clean up the column names. `nrow()` is equal to the number of rows in our object. If we were to change to 130 simulations, that would generate 130 rows, and `nrow()` would be equal to 130.

```
monte_carlo_sim_51 <-
  monte_carlo_sim_51 %>%
  mutate(month = seq(1:nrow(.))) %>%
  select(month, everything()) %>%
  `colnames<-`(c("month", names(starts))) %>%
  mutate_all(funs(round(., 2)))

tail(monte_carlo_sim_51 %>%  select(month, sim1, sim2,
                                    sim49, sim50), 3)
```

```
# A tibble: 3 x 5
  month  sim1  sim2 sim49 sim50
  <dbl> <dbl> <dbl> <dbl> <dbl>
1  119.  1.77  3.12  2.85  2.51
```

```
2   120.   1.73   3.17   2.90   2.52
3   121.   1.82   3.26   2.84   2.59
```

We have accomplished our goal of running 51 simulations and could head to data visualization now, but let's explore an alternative method using the the rerun() function from purrr. As its name imples, this function will "rerun" another function and we stipulate how many times to do that by setting .n = number of times to rerun. For example to run the simulation_accum_1 function 5 times, we would set the following:

```
monte_carlo_rerun_5 <-
    rerun(.n = 5,
        simulation_accum_1(1,
                           120,
                           mean_port_return,
                           stddev_port_return))
```

That returned a list of 5 data frames, or 5 simulations. We can look at the first few rows of each data frame by using map(..., head).

```
map(monte_carlo_rerun_5, head)
```

```
[[1]]
# A tibble: 6 x 1
  growth
   <dbl>
1   1.00
2   0.981
3   1.01
4   1.05
5   1.04
6   1.03

[[2]]
# A tibble: 6 x 1
  growth
   <dbl>
1    1.00
2    1.01
3    1.04
4    1.05
5    1.04
6    1.07

[[3]]
```

```
# A tibble: 6 x 1
  growth
  <dbl>
1   1.00
2   1.03
3   1.05
4   1.07
5   1.08
6   1.09

[[4]]
# A tibble: 6 x 1
  growth
  <dbl>
1   1.00
2   0.988
3   0.972
4   0.964
5   0.971
6   0.964

[[5]]
# A tibble: 6 x 1
  growth
  <dbl>
1   1.00
2   1.02
3   1.04
4   0.995
5   0.949
6   0.970
```

Let's consolidate that list of data frames to one `tibble`. We start by collapsing to vectors with `simplify_all()`, then add nicer names with the `names()` function and finally coerce to a `tibble` with `as_tibble()`. Let's run it 51 times to match our previous results.

```
reruns <- 51

monte_carlo_rerun_51 <-
rerun(.n = reruns,
      simulation_accum_1(1,
                         120,
                         mean_port_return,
                         stddev_port_return)) %>%
```

```
simplify_all() %>%
`names<-`(paste("sim", 1:reruns, sep = " ")) %>%
as_tibble() %>%
mutate(month = seq(1:nrow(.))) %>%
select(month, everything())

tail(monte_carlo_rerun_51 %>%
        select(`sim 1`, `sim 2`,
                `sim 49`, `sim 50`), 3)
```

```
# A tibble: 3 x 4
  `sim 1` `sim 2` `sim 49` `sim 50`
    <dbl>   <dbl>    <dbl>    <dbl>
1    2.95    2.92     2.17     1.61
2    2.92    2.99     2.23     1.57
3    2.91    2.99     2.26     1.62
```

Now we have two objects holding the results of 51 simulations, monte_carlo_rerun_51 and monte_carlo_sim_51.

Each has 51 columns of simulations and 1 column of months. Note that we have 121 rows because we started with an intitial value of 1, and then simulated returns over 120 months.

11.4   Visualizing Simulations with ggplot

Now let's get to ggplot() and visualize the results in monte_carlo_sim_51. The same code flows for visualization would also apply to monte_carlo_rerun_51 but we will run them for only monte_carlo_sim_51.

We start with a chart of all 51 simulations and assign a different color to each one by setting ggplot(aes(x = month, y = growth, color = sim)). ggplot() will automatically generate a legend for all 51 time series but that gets quite crowded. We will suppress the legend with theme(legend.position = "none").

```
monte_carlo_sim_51 %>%
  gather(sim, growth, -month) %>%
  group_by(sim) %>%
  ggplot(aes(x = month, y = growth, color = sim)) +
```

```
geom_line() +
theme(legend.position="none")
```

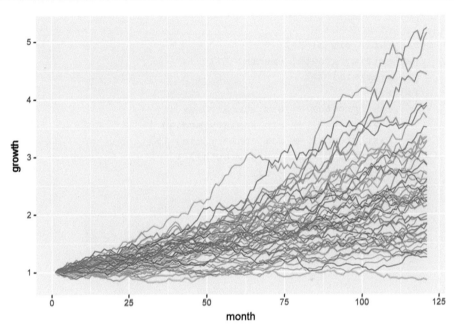

FIGURE 11.1: 51 Simulations ggplot

Figure 11.1 gives a good sense for how our 51 simulations are distributed.

We can check the minimum, maximum and median simulation with the summarise() function here.

```
sim_summary <-
monte_carlo_sim_51 %>%
  gather(sim, growth, -month) %>%
  group_by(sim) %>%
  summarise(final = last(growth)) %>%
  summarise(
          max = max(final),
          min = min(final),
          median = median(final))
sim_summary


# A tibble: 1 x 3
    max    min median
```

```
<dbl> <dbl>  <dbl>
1  5.23 0.860   2.29
```

We then tweak our original visualization by including only the maximum, minimum and median simulation result.

```
monte_carlo_sim_51 %>%
  gather(sim, growth, -month) %>%
  group_by(sim) %>%
  filter(
      last(growth) == sim_summary$max ||
      last(growth) == sim_summary$median ||
      last(growth) == sim_summary$min) %>%
  ggplot(aes(x = month, y = growth)) +
  geom_line(aes(color = sim))
```

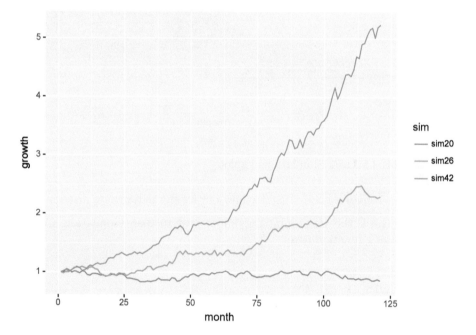

FIGURE 11.2: Min, Max, Median Sims ggplot

Figure 11.2 is cleaner and still provides an undertanding of the distribution.

Since we have a wide range of possible end values, let's examine the quantiles for those values.

First, we assign different probability values to a vector with **probs** <- **c(.005,** **.025, .25, .5, .75, .975, .995).**

```
probs <- c(.005, .025, .25, .5, .75, .975, .995)
```

Next we want to isolate the end values for our 51 dollar growth simulations.
We use `summarise(final = last(growth))` and create a new object called
`sim_final_quantile` to hold final values.

```
sim_final_quantile <-
monte_carlo_sim_51 %>%
  gather(sim, growth, -month) %>%
  group_by(sim) %>%
  summarise(final = last(growth))
```

Finally we call the `quantile()` function on `sim_final_quantile$final` and
pass in our vector of probability values, which we labeled `probs`.

```
quantiles <-
  quantile(sim_final_quantile$final, probs = probs) %>%
  tibble() %>%
  `colnames<-`("value") %>%
  mutate(probs = probs) %>%
  spread(probs, value)

quantiles[, 1:6]
```

```
# A tibble: 1 x 6
  `0.005` `0.025` `0.25` `0.5` `0.75` `0.975`
    <dbl>   <dbl>  <dbl> <dbl>  <dbl>   <dbl>
1   0.958    1.26   1.77  2.29   2.94    4.97
```

The results tell us the exact quantile cut points for the simulations.

11.5 Visualizing Simulations with highcharter

Before we get to Shiny, let's port our simulation work over to `highcharter`
but experiment using a tidy data frame instead of an `xts`. To convert to a tidy
`tibble`, we use the `gather()` function from `tidyr`.

```
mc_gathered <-
  monte_carlo_sim_51 %>%
```

```
gather(sim, growth, -month) %>%
group_by(sim)
```

In a departure from our normal code flow, we will now pass this `tibble` directly
to the `hchart()` function, specify the type of chart as `line` and then work
with a similar grammar to `ggplot()`. The difference is we use `hcaes`, which
stands for `highcharter aesthetic`, instead of `aes`.

```
# This takes a few seconds to run
hchart(mc_gathered,
       type = 'line',
       hcaes(y = growth,
             x = month,
             group = sim)) %>%
  hc_title(text = "51 Simulations") %>%
  hc_xAxis(title = list(text = "months")) %>%
  hc_yAxis(title = list(text = "dollar growth"),
           labels = list(format = "${value}")) %>%
  hc_add_theme(hc_theme_flat()) %>%
  hc_exporting(enabled = TRUE) %>%
  hc_legend(enabled = FALSE)
```

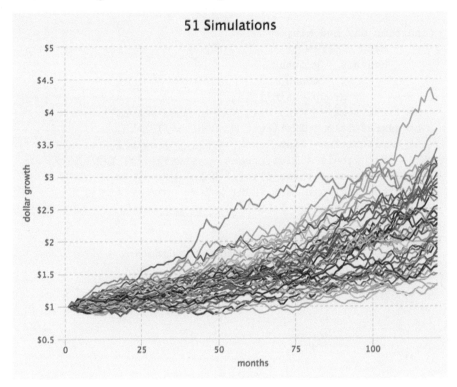

FIGURE 11.3: Monte Carlo Simulations highcharter

In Figure 11.3, we plotted 51 lines in `highcharter` using a tidy `tibble`. That's nice because when we go to Shiny, the end user will be able to choose any number of simulations. It's also nice for us since we now can work with the same data logic for both `ggplot()` and `highcharter`.

Very similar to what we did with `ggplot`, let's isolate the maximum, minimum and median simulations and save them to an object called `mc_max_med_min`.

```
mc_max_med_min <-
  mc_gathered %>%
  filter(
      last(growth) == sim_summary$max ||
      last(growth) == sim_summary$median ||
      last(growth) == sim_summary$min) %>%
  group_by(sim)
```

Now we pass that filtered object to `hchart()`.

```
hchart(mc_max_med_min,
       type = 'line',
       hcaes(y = growth,
             x = month,
             group = sim)) %>%
  hc_title(text = "Min, Max, Median Simulations") %>%
  hc_xAxis(title = list(text = "months")) %>%
  hc_yAxis(title = list(text = "dollar growth"),
           labels = list(format = "${value}")) %>%
  hc_add_theme(hc_theme_flat()) %>%
  hc_exporting(enabled = TRUE) %>%
  hc_legend(enabled = FALSE)
```

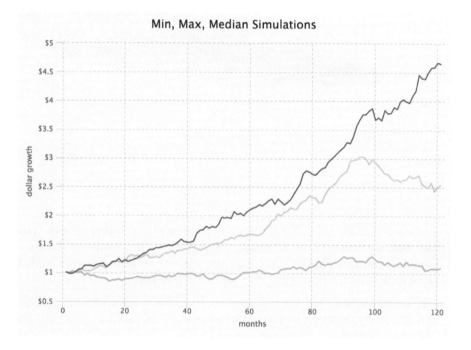

FIGURE 11.4: Monte Carlo Min Max Median highcharter

Figure 11.4 shows only three lines but still gives an nice picture of our simulation results.

11.6   Shiny App Monte Carlo

Now to our Shiny app wherein a user can build a custom portfolio and then choose a number of simulations to run and a number of months to simulate into the future.

Have a look at the app in Figure 11.5:

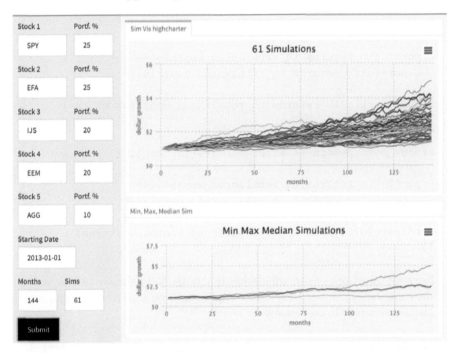

FIGURE   11.5:   www.reproduciblefinance.com/shiny/monte-carlo-simulation/

The input sidebar is our usual except we also ask for the number of months to be simulated and the number of simulations to be run. The code to create those `numericInput()` fields is below.

```
fluidRow(
  column(5,
  numericInput("sim_months", "Months", 120,
               min = 6, max = 240, step = 6)),
  column(5,
```

```
numericInput("sims", "Sims", 51,
             min = 31, max = 101, step = 10))
)
```

From here, we calculate portfolio returns and save them as portfolio_returns_tq_rebalanced_monthly, and then find the mean and standard deviation of those returns. Those are the parameters we need for the simulation.

```
mean_port_return <- eventReactive(input$go, {

  portfolio_returns_tq_rebalanced_monthly <-
    portfolio_returns_tq_rebalanced_monthly()

  mean(portfolio_returns_tq_rebalanced_monthly$returns)
})

stddev_port_return <- eventReactive(input$go, {

  portfolio_returns_tq_rebalanced_monthly <-
    portfolio_returns_tq_rebalanced_monthly()

  sd(portfolio_returns_tq_rebalanced_monthly$returns)
})
```

Next we insert one of our simulation functions. Very similar to how we coded the contribution to volatility Shiny application, we insert our custom function just as we wrote it previously.

```
simulation_accum_1 <- function(init_value, N, mean, stdev) {
    tibble(c(init_value, 1 + rnorm(N, mean, stdev))) %>%
    `colnames<-`("returns") %>%
    mutate(growth =
            accumulate(returns, function(x, y) x * y)) %>%
    select(growth)

}
```

Then, we call eventReactive() to run the simulation, following the same logic as we did above.

```
sims <- eventReactive(input$go, {input$sims})
```

```
monte_carlo_sim <- eventReactive(input$go, {

  sims <- sims()

  starts <-
    rep(1, sims) %>%
    set_names(paste("sim", 1:sims, sep = ""))

  map_dfc(starts, simulation_accum_1,
          N = input$sim_months, mean = mean_port_return(),
          stdev = stddev_port_return()) %>%
  mutate(month = seq(1:nrow(.))) %>%
  select(month, everything()) %>%
  `colnames<-`(c("month", names(starts))) %>%
  gather(sim, growth, -month) %>%
  group_by(sim) %>%
  mutate_all(funs(round(., 2)))

})
```

We now have a reactive object called `monte_carlo_sim()` which holds our 51 simulations of the custom portfolio. We can visualize with `highcharter()`, exactly as we did in the visualization section.

```
renderHighchart(
  hchart( monte_carlo_sim(),
          type = 'line',
          hcaes(y = growth,
                x = month,
                group = sim)) %>%
  hc_title(text = paste(sims(),
                        "Simulations",
                        sep = " ")) %>%
  hc_xAxis(title = list(text = "months")) %>%
  hc_yAxis(title = list(text = "dollar growth"),
           labels = list(format = "${value}")) %>%
  hc_add_theme(hc_theme_flat()) %>%
  hc_exporting(enabled = TRUE) %>%
  hc_legend(enabled = FALSE)
)
```

And, finally, we isolate the minimum, median and maximum simulations.

```r
renderHighchart({

sim_summary <-
  monte_carlo_sim() %>%
  summarise(final = last(growth)) %>%
  summarise(
            max = max(final),
            min = min(final),
            median = median(final))

mc_max_med_min <-
  monte_carlo_sim() %>%
  filter(
      last(growth) == sim_summary$max ||
      last(growth) == sim_summary$median ||
      last(growth) == sim_summary$min)

  hchart(mc_max_med_min,
         type = 'line',
         hcaes(y = growth,
               x = month,
               group = sim)) %>%
  hc_title(text = "Min Max Median Simulations") %>%
  hc_xAxis(title = list(text = "months")) %>%
  hc_yAxis(title = list(text = "dollar growth"),
           labels = list(format = "${value}")) %>%
  hc_add_theme(hc_theme_flat()) %>%
  hc_exporting(enabled = TRUE) %>%
  hc_legend(enabled = FALSE)
})
```

That wraps our last Shiny dashboard and I concluded with this one because it shows a powerful use for Shiny: we could have written any simulation function we wanted and then displayed the results. The end user sees the charts and we, as the R coders, have total freedom to build more powerful functions and then deliver them via Shiny.

Concluding Practice Applications

Our work on asset contribution to portfolio standard deviation and Monte Carlo simulation is complete. We started with step-by-step matrix algebra, then created two custom functions to calculate contribution to standard deviation, and then deployed to Shiny.

Next we ran a simulation, created our own function, and deployed to Shiny. Both use cases demonstrate the power at our fingertips when writing custom functions, both for running analyses and delivering results to end users via Shiny. Even though these are a little more complex than our previous work, we maintained reproducible and reusable code that can be applied to whatever tasks wait for us in the future.

If you are starting a new R session and wish to run the code for practice and applications, first get the portfolio returns objects:

www.reproduciblefinance.com/code/get-returns/

And then see these resources:

www.reproduciblefinance.com/code/component-contribution/

www.reproduciblefinance.com/code/monte-carlo/

Appendix: Further Reading

Below is some suggested reading on the topics covered in this book.

Portfolio Theory and Financial Modelling

For readers who want to explore those concepts generally, *Investments* by Bodie, Kane and Marcus is a good place to start. For readers who want to delve deeper into financial modeling and optimization with R, I recommend *Financial Risk Modelling and Portfolio Optimization with R* by Bernhard Pfaff or *Financial Analytics with R Building a Laptop Laboratory for Data Science* by Bennett and Hugen.

R programming

For readers who want to learn R, please start with *R for Data Science* by Garrett Grolemund and Hadley Wickham.

R Technology Tidbits

Below are some brief resource pointers for some of the base technologies that were used to create this book and the Shiny applications.

RMarkdown

RMarkdown is a file framework for creating reproducible reports (PDF and HTML), notebooks, books and more.

This entire book was written using RMarkdown and it demonstrates how smooth it is to include plain text prose (like these words), R code and the results of R code in a document.

See:

rmarkdown.rstudio.com/articles_intro.html

r4ds.had.co.nz/r-markdown.html

for introductions to RMarkdown.

The pipe operator

The %>% operator appears in most of the code in this book. It looks like this.

```
returns %>%
  head()
```

We would read that code chunk as, "start with the `returns` data *and then* print just the head of that data". The `%>%` operator is the 'and then'. We start with a data object *and then* apply a function, *and then* apply another, *and then* select a column until we are finished. We do not need to save the variable at each step, which makes the code more readable and less duplicative.[1].

For more background, see Chapter 18 of *R For Data Science*:

r4ds.had.co.nz/pipes.html

Shiny

Shiny is an R package that allows you to turn your R code into an interactive web application, which is called a Shiny application. With Shiny, R coders can create interactive dashboards, that use HTML, CSS and JavaScript, without having to learn those languages.

If you wish to delve deeper into Shiny and best practices, there are free courses and materials available at:

shiny.rstudio.com/tutorial/ and

datacamp.com/courses/building-web-applications-in-r-with-shiny

Flexdashboard

Flexdashboard is a templating R package for building Shiny dashboards with RMarkdown. All of the Shiny applications in this book are built using this template. You can learn more about how to insert charts into your dashboard at the `flexdashboard` homepage.

See here for more information:

rmarkdown.rstudio.com/flexdashboard/index.html

Packages and Github

This book did not cover two important pieces of the technology stack: building packages and version control with github.

`PerformanceAnalytics`, `dplyr`, `tidyquant` - they are all packages and that means it's easy for us to access their functions and data once we have installed them. Well, we could also save all of our own work into a package, so that others on our team can easily access our functions and data. Saving our work and data into R packages is a great way to share them for reuse and reproduction. If you work at a financial institution, the best and most common practice is to save your functions and models into private packages in private repositories. The reason I did not cover package building is that Hadley Wickham has already covered it in his book *R Packages*, which is available for free here:

r-pkgs.had.co.nz/

---

[1]For the curious, see here: www.reproduciblefinance.com/code/standard-deviation-by-hand/

Version control with github (or another technology) is also hugely important. It's how we and our team keep track of changes to our code, models, data and packages. Why is that crucial? If our team or a PM is using our model and we change the original code, how will other people get the new version? How will they even know about the new version? The answer is a version control system. In brief, the version control repository is where we store our code versions.

This topic has been covered very well by Jenny Bryan at the web project:

happygitwithr.com

Index